Industry 4.0, AI, and Data Science

Demystifying Technologies for Computational Excellence: Moving Towards Society 5.0
Series Editors: Vikram Bali and Vishal Bhatnagar

This series encompasses research work in the field of Data Science, Edge Computing, Deep Learning, Distributed Ledger Technology, Extended Reality, Quantum Computing, Artificial Intelligence, and various other related areas, such as natural-language processing and technologies, high-level computer vision, cognitive robotics, automated reasoning, multivalent systems, symbolic learning theories and practice, knowledge representation and the semantic web, intelligent tutoring systems, AI and education.

The prime reason for developing and growing out this new book series is to focus on the latest technological advancements – their impact on the society, the challenges faced in implementation, and the drawbacks or reverse impact on the society due to technological innovations. With the technological advancements, every individual has personalized access to all the services, all devices connected with each other communicating amongst themselves, thanks to the technology for making our life simpler and easier. These aspects will help us to overcome the drawbacks of the existing systems and help in building new systems with latest technologies that will help the society in various ways proving Society 5.0 as one of the biggest revolutions in this era.

Industry 4.0, AI, and Data Science
Research Trends and Challenges
Edited by Vikram Bali, Kakoli Banerjee, Narendra Kumar, Sanjay Gour, and Sunil Kumar Chawla

For more information on this series, please visit: www.routledge.com/Demystifying-Technologies-for-Computational-Excellence-Moving-Towards-Society-5.0/book-series/CRCDTCEMTS

Industry 4.0, AI, and Data Science
Research Trends and Challenges

Edited by
Vikram Bali, Kakoli Banerjee, Narendra Kumar, Sanjay Gour,
and Sunil Kumar Chawla

CRC Press
Taylor & Francis Group
Boca Raton London New York

CRC Press is an imprint of the
Taylor & Francis Group, an **informa** business

First edition published 2022
by CRC Press
6000 Broken Sound Parkway NW, Suite 300, Boca Raton, FL 33487-2742

and by CRC Press
2 Park Square, Milton Park, Abingdon, Oxon OX14 4RN

CRC Press is an imprint of Taylor & Francis Group, LLC

Library of Congress Cataloging-in-Publication Data
Names: Bali, Vikram, editor.
Title: Industry 4.0, AI, and data science: research trends and
challenges / edited by Vikram Bali, Kakoli Banerjee,
Narendra Kumar, Sanjay Gour, and Sunil Kumar Chawla.
Description: First edition. | Boca Raton, FL: CRC Press, 2021. |
Series: Demystifying technologies for computational excellence |
Includes bibliographical references and index.
Identifiers: LCCN 2021004328 (print) | LCCN 2021004329 (ebook) |
ISBN 9780367562854 (hardback) | ISBN 9780367562915 (paperback) |
ISBN 9781003097181 (ebook)
Subjects: LCSH: Industry 4.0.
Classification: LCC T59.6 .I34 2021 (print) |
LCC T59.6 (ebook) | DDC 658.4/038028563–dc23
LC record available at https://lccn.loc.gov/2021004328
LC ebook record available at https://lccn.loc.gov/2021004329

ISBN: 978-0-367-56285-4 (hbk)
ISBN: 978-0-367-56291-5 (pbk)
ISBN: 978-1-003-09718-1 (ebk)

Typeset in Times
by Newgen Publishing UK

Contents

Contributors

Aditi
Department of Computer Science &
   Engineering
FET, PDM University
Bahadurgarh, Haryana, India

Shivani Bali
Jaipuria Institute of Management
Noida, Uttar Pradesh, India

Vikram Bali
JSS Academy of Technical Education
Noida, Uttar Pradesh, India

Vishal Bhatnagar
Netaji Subhas University of Technology
East Campus
Delhi, India

Suryansh Chauhan
Raja Balwant Singh Engineering
   Technical Campus
Agra, Uttar Pradesh, India

Akhilesh Kumar Choudhary
BSNL
Ghaziabad, Uttar Pradesh, India

Pushpa Choudhary
G L Bajaj Institute of Technology and
   Management
Greater Noida, Uttar Pradesh, India

Tripti Dua
Jaipur Engineering College and
   Research Centre
Jaipur, Rajasthan, India

Aman Dureja
Department of Computer Science &
   Engineering
FET, PDM University
Bahadurgarh, Haryana, India

Vikas Goel
Department of IT, KIET Group of
   Institutions
Ghaziabad, Uttar Pradesh India

Sanjay Gour
Jaipur Engineering College and
   Research Centre
Jaipur, Rajasthan, India

Amit Kumar Gupta
Department of CA
KIET Group of Institutions
Ghaziabad, Uttar Pradesh, India

Mansi Gupta
Lal Bahadur Shastri Institute of
   Management
Dwarka, New Delhi, India

Yamini Gupta
JSS Academy of Technical Education
Noida, Uttar Pradesh, India

Anshuman Jaiswal
Raja Balwant Singh Engineering
   Technical Campus
Agra, Uttar Pradesh, India

Sapna Juneja
IMS Engineering College
Ghaziabad, Uttar Pradesh, India

Abhinav Juneja
KIET Group of Institutions
Ghaziabad, Uttar Pradesh, India

Mahabaleshwar S. Kakkasageri
Basaveshwar Engineering College
   (Autonomous)
Bagalkot, Karnataka, India

Amandeep Kaur
Punjab Engineering College (Deemed to
   be University)
Chandigarh, Punjab, India

P.K. Khosla
C-DAC
Mohali, Punjab, India

Akshay Kumar
Raja Balwant Singh Engineering
  Technical Campus
Agra, Uttar Pradesh, India

Narendra Kumar
Dean, BlueCrest University
Liberia, West Africa

Sachin Kumar
Department of CSE
Ajay Kumar Garg Engineering College
Ghaziabad, Uttar Pradesh, India

Sangeeta Mangesh
JSS Academy of Technical Education
Noida, Uttar Pradesh, India

Sunil Kumar S. Manvi
Reva University
Bangalore, Karnataka, India

Arvind Maurya
Technology Director
HCL Technologies
Noida, Uttar Pradesh, India

Sneha Mishra
Galgotias University
Greater Noida, Uttar Pradesh, India

Neerja Mittal
Central Scientific Instruments
  Organisation
Chandigarh, Punjab, India

Jacob Muchuchuti
Botswana Accountancy College
Gaborone, Botswana

Stewart Muchuchuti
Botswana Accountancy College
Gaborone, Botswana

Julius Onyancha
University of Sunderland
Sunderland, UK

Payal Pahwa
Department of Computer Science &
  Engineering
BPIT
Rohini, New Delhi, India

Arvind Panwar
University School of Information
Communication and Technology Guru
  Gobind Singh Indraprastha University
Dwarka, Delhi, India

Valentina Plekhanova
University of Sunderland, UK

Priya Porwal
Galgotias University
Greater Noida, Uttar Pradesh, India

Anju Rajput
Jaipur Engineering College and
  Research Centre
Jaipur, Rajasthan, India

Mamata J. Sataraddi
Basaveshwar Engineering College
  (Autonomous)
Bagalkot, Karnataka, India

Vishesh Saxena
JSS Academy of Technical Education,
Noida, Uttar Pradesh, India

Sonali Semwal
Lal Bahadur Shastri Institute of
  Management
Delhi, India

Lavkush Sharma
Raja Balwant Singh Engineering
  Technical Campus
Agra, Uttar Pradesh, India

Krishna Kant Sharma
Expert Member
Ministry of Communications
Government of India

Arjun Singh
GL Bajaj Institute of Technology and
  Management
Greater Noida, Uttar Pradesh, India

Arun Kumar Singh
GL Bajaj Institute of Technology and
  Management
Greater Noida, Uttar Pradesh, India

Reema Thareja
Assistant Professor Department of
  Computer Science
Shyama Prasad Mukherji College
Delhi University
Delhi, India

Utkarsh Tripathi
JSS Academy of Technical Education
Noida, Uttar Pradesh, India

Hemant Upadhyay
BMIET
Sonepat, Haryana, Delhi

Dileep Kumar Yadav
Galgotias University
Greater Noida, Uttar Pradesh, India

Preface

Greetings!

The Spreadsheet era is over. A Google search, a passport check, a tweet, your history of online shopping. They all contain information that can be stored, analyzed, and monetized. In real time, supercomputers and algorithms allow us to make sense of a growing amount of knowledge. CPUs are expected to exceed the human brain's processing capacity in less than ten years. Many CEOs, CTOs, and decision makers of companies are dreaming of ways to reinvent their business with the advent of big data and fast computing capacity. When they want to launch a new product or service, they look at data analytics for consumer insights, demand, demographic targeting, etc. Artificial intelligence and data science are increasingly being implemented into the enterprise.

This book calls for insight into Industry 4.0.-based data science and artificial intelligence techniques. This book bridges the gap between innovations and strategic tactics involved in data science and artificial intelligence techniques. Machine learning and data science are now becoming an important part of a variety of industrial and academic studies.

In Chapter 1, "Predicting Fraudulent Motor Vehicle Insurance Claims Using Data Mining Model," it is elaborated that in recent years, insurance fraud detection has attracted a great deal of concern and attention as the insurance industry has witnessed an increase in the number of fraudulent claims. The aim of this research was to develop a data mining model that would predict fraudulent insurance claims. The research also seeks to establish the variables with the most predictive power. From a data set with 1,000 claims, 70 percent were used for training the machine using Python Programming Language and the remainder being used to test the accuracy of the algorithms used. Twenty-nine attributes were observed to have influence in predicting the potentially fraudulent behavior of the claims and variables such as policyholder's hobbies and the extent of damage on the motor vehicle were found to have the most predictive influence. As for the predictive influence, the Decision Tree had the highest accuracy level compared to the other algorithms.

In Chapter 2, "Novel 8:1 Multiplexer for Low Power and Area Efficient Design in Industry 4.0," it is elaborated that we are in the midst of digitization of manufacturing. Industry 4.0 is the fourth revolution, which occurred in the field of manufacturing. This fourth industrial revolution is accommodating smart and autonomous systems fuelled by data and machine-learning. Industry 4.0 is the optimization of computerization, which was introduced in Industry 3.0. Now apart from computerization, decision making without human involvement is more focused criteria. Now focus is more on developing smart machines, which can access more data but without consuming large space. Therefore, it is necessary to develop area efficient machines so that our industry will become more efficient and productive and less wasteful. As Industry 4.0 is digital transformation of manufacturing and processes, multiplexer design with relatively less number of transistors results in miniature products.

In Chapter 3, "Data Science and AI for E-Governance: A Step towards Society 5.0," it is elaborated that the outbreak of the pandemic Covid-19 has hit the world in quite a horrific manner; where almost all the governments are struggling hard to ensure the safety and security of citizens as well as maintaining the country's economic status. In such times, the benefits of data science and artificial analysis can be explored. The tracking of Covid-19 patients, source of transmission, and future prospects has been possible with the help of data; which is an excellent example of data science in the present times. Open data availability on platforms like Kaggle provides an opportunity to researchers to develop various solutions and improve upon them. Chatty Gargoyle at Denver International Airport provides a glimpse into the world of artificial intelligence. This chapter aims to provide a detailed study of various data science and AI applications for E-government and societal benefits. The concepts of data science and AI are discussed in detail along with the programming tools which can be used to build solutions out of these as well as integrate these domains. Further, the existing applications of these paradigms in the said areas have been summarized. The future prospects and trends are also discussed in detail to provide research directions in these domains.

In Chapter 4, "Application Areas of Data Science and AI for Improved Society 5.0 Era," it is elaborated that the grooming technologies in this new era are data science and artificial intelligence serving their best to society in multidiscipline to improve human lives. Areas like medical science, agriculture, banking, communication, retail, social media, search engines, autonomous vehicles, and various industries are training their employees to identify patterns and gather insights from large and complex datasets. The world is surrounded by massive amount of data and it increases exponentially. Data science is the responsibility to specialize the source of data, the relevancy of that data, and how to dig out the useful information from that data. Any data can be seen from various perspectives and thus data science has become the biggest asset nowadays.

In Chapter 5, "Applying Machine-Learning and Internet of Things in Healthcare," it is elaborated that the Internet of Things (IoT) advancement has pulled in much thought of late for its capacity to facilitate the strain on social protection structures realized by a developing masses and a climb in perpetual sickness. Standardization is a key issue obliging progression around there, and thus this examination presents a quality prototype for applications in upcoming IoT therapeutic administrations structures. This chapter shows the top tier look to each locale of the prototype, surveying its characteristics, shortcomings, and as the rule fittingness for IoT social protection unit. Troubles faced by social protection of IoT faces, including safety issues, assurance, wearable capabilities, and not proper utilization of action, are shown and suggestions are provided for future research headings.

In Chapter 6, "Artificial Intelligence: The New Expert in Medical Treatment," it is elaborated that artificial intelligence is a computer algorithm's ability to simulate observations without direct human oversight. A fast-growing use of AI in the medical field of photos – an area that relies on deep learning, a sophisticated and efficient type of data science where a set of labeled data are fed into algorithms that detect characteristics among them and understand how to identify various patterns. This strategy has shown to be successful and beneficial in diagnostic testing from cancer to eye disorders. With artificial intelligence growing rapidly in every field, the world has entered the fourth industrial revolution, and the organizations that are enthusiastic about adjusting and embracing new technology and procedures will turn out to be the industry leaders of tomorrow. One of the major obstacles for artificial intelligence in the field of healthcare is simply the aversion to change. For centuries only doctors held exclusive knowledge and issued treatment procedures but the use of artificial intelligence in the future can cause a shift in culture of interactions between doctors and patients.

In Chapter 7, "Machine Learning Approach for Breast Cancer Early Diagnosis," it is elaborated that machine learning has been the most recommended approach for predicting the early detection of breast cancer. With the availability of datasets that have details about the features extracted from the mammograms as well as other imaging techniques, the modeling and training can aid in the early detection of breast cancer. Such early diagnostic strategies can focus on providing timely access to cancer treatment thereby improving the quality of life of these cancer patients.

In Chapter 8, "Intelligent Surveillance System Using Machine Learning," it is elaborated that in today's era we are facing major security issues; consequently, we need several specially trained personnel to attain the desired security. The development in the field of facial recognition plays an important role in today's world for crime prevention. Our proposed solution to the aforementioned matter is a surveillance system that has capability of facial recognition, which can detect intruders to restricted or high-security areas found at a specific location under the surveillance of a CCTV camera.

In Chapter 9, "Cyber Security: An Approach to Secure IoT from Cyber Attacks Using Deep Learning," it is elaborated that as the popularity and usage of Internet of Things (IoT) devices is increasing day by day in today's world, the security of these IoT devices from cyber attacks has become a major concern for all the stakeholders. The main purpose of providing security to IoT devices is to secure the connecting devices and available networks in the Internet of things. The scope and role of IoT driven devices and agents are exploding at a very fast pace. Industry 4.0 is evolving and it mainly relies on the incorporation of IoT driven technology for its growth. There is a need to establish trust and security in use of IoT so as to make these things available to all stakeholders with different application domains.

In Chapter 10, "Learning the Dynamic Change of User Interests from Noise Web Data," it is elaborated that the web is noise, inconsistent and irrelevant by nature, finding useful information that defines interest of the user has become a challenge. Noisy web data is currently considered as data that is not part of the main web page content. However, not every type of information that forms part of the main web content meets the interests of web users. Existing research acknowledges that there is a need to propose machine learning tools capable of addressing problems with data available on the web and what users are interested in. This research work presents an investigation of current research work proposed in minimizing the levels of noisy data on the web. It examines their contribution as well as existing challenges. It proposes an approach that learns web data in relation to the interest of users. The proposed approach considers the dynamic change of user interest as well as evolving web data.

In Chapter 11, "Artificial Intelligence Techniques Based Routing Protocols in VANETs: A Review," it is elaborated that the Vehicle Ad Hoc Network (VANET) is a type of wireless network consisting of vehicles and roadside communication devices. As VANET is associated with life-critical applications, it is important to consider each component of VANET design, functionality, software, and problems before introducing them. The important goal of VANET is to offer seamless communication for people traveling on the road to collect and relay message from every nearby vehicles in the event of urgent situations such as severe traffic jams, collisions, lane shift, speed limit, hazard or road condition alerts, position alert services, and in the event of climatic disasters, etc. Due to the high mobility of the vehicles, frequent network disconnection happens and many problems emerge in VANETs such as routing, synchronization, network congestion, network control, information management, privacy, and security, etc. This chapter reviews the current state of the research on recent artificial intelligence techniques based VANET routing. The chapter also addresses on-going research works on the usage of artificial intelligence techniques for routing in VANET and future challenges need to be addressed in providing the artificial intelligence techniques for routing in VANETs.

In Chapter 12, "A Comparison of Different Consensus Protocols: The Backbone of the Blockchain Technology," it is elaborated that after launching bitcoin, Blockchain is gaining massive growth in the last decade. Researchers are trying to find blockchain potential in other areas, also such as healthcare, government services, data access, and many more. As all we knew, whenever technology is expanding exponentially, it is crucial to discuss all the critical components of technology. Some key elements of blockchain technology are cryptography, decentralization, P2P network, validity, consensus protocol, and distributed ledger. In this chapter the author focuses on the consensus algorithm. A consensus algorithm is a backbone and primary root of revolutionary technology. A consensus algorithm creates an agreement that decides how a block appends as a new block. The author discusses two types of algorithms, the first proof-based and the second voting-based. At the end of this chapter, we present a detailed comparison of the blockchain consensus algorithm.

In Chapter 13, "Blockchain in AI: Review of Decentralized Smart System," it is elaborated that in recent times, the two technologies – artificial intelligence (AI) and blockchain – are the most trending and disruptive in all the technologies. Undoubtedly both technologies are covering all the areas of engineering and science at an exceptional rate. These technologies offer various degrees of technological complexity and multi-dimensional business implications.

In Chapter 14, "Financial Portfolio Optimization: an AI Based Decision-Making Approach," it is elaborated that in financial data science and more specifically in investment analytics, portfolio optimization is a very important aspect. A portfolio consists of multiple securities, each having its own weight. Based upon these weights, the overall Portfolio Return and Risk are determined. Investors try to find the optimal portfolio that helps in maximizing return for a given risk by adjusting these weights given to securities. In this chapter, a diverse and exemplar portfolio is considered.

It has stocks from four companies belonging to four different Sectors (from Indian market NSE Index): Hindustan Unilever, Reliance Industries, Tata Consultancy Services (TCS), and Sun Pharma.

In Chapter 15, "Intelligent Framework and Metrics for Assessment of Smart Cities," it is elaborated that the world population has crossed 7.8 billion and is expected to touch 10 billion by 2050. Rapid urbanization and population growth have put all resources under tremendous pressure. Technical evolutions such as Digitalization and Internet of Things are increasingly being used to make smart cities which are more pervasive, efficient, and sustainable. Traditional literature has proposed various frameworks for smart cities but the recent Covid-19 pandemic has demonstrated the glaring shortcomings in existing smart city framework. This chapter proposes a smart city framework which is more inclusive, more adaptive, more dynamic, and more sustainable. It attempts to bridge the gap between the existing literature and the challenges of tomorrow.

We wish all our readers and their family members good health and prosperity.

Dr. Vikram Bali
Dr. Kakoli Banerjee
Dr. Narendra Kumar
Dr. Sanjay Gour
Mr. Sunil Kumar Chawla

Editors

Vikram Bali is professor and head of the Computer Science and Engineering Department at JSS Academy of Technical Education, Noida, India. He graduated from REC, Kurukshetra – B.Tech (CSE), Post Graduation from NITTTR, Chandigarh – M.E (CSE) and Doctorate (Ph.D) from Banasthali Vidyapith, Rajasthan. He has more than 20 years of rich academic and administrative experience. He has published more than 50 research papers in international journals/conferences, has edited books, and authored five text books. He also has published five patents. He is on the editorial board and on the review panel of many international journals. He is series editor for three book series of CRC Press, Taylor & Francis Group. He is a lifetime member of IEEE, Indian Society for Technical Education (ISTE), Computer Society of India (CSI) and Institution of Engineers (IE). He was awarded the Green Thinker Z-Distinguished Educator Award 2018 for remarkable contribution in the field of Computer Science and Engineering at the Third International Convention on Interdisciplinary Research for Sustainable Development (IRSD) at the Confederation of Indian Industry (CII), Chandigarh. He has also attended the Faculty Enablement program organized by Infosys and NASSCOM. He has been the member of board of studies of different Indian Universities and member of organizing committee for various National and International Seminars/Conferences. He is working on four sponsored research projects funded by TEQIP-3 and Unnat Bharat Abhiyaan. His research interest includes Software Engineering, Cyber Security, Automata Theory, CBSS, and ERP.

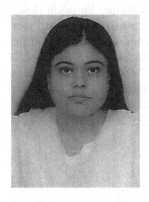

Kakoli Banerjee has been associate professor in Computer Science and Engineering Department currently serving JSS Academy of Technical Education, Noida, India for the last 17 years. She graduated from IET, Kanpur – B.Tech (CSE), did her Post Graduation from MNNIT, Allahabad – M.Tech (CSE) and her Doctorate (Ph.D) from Shobhit University, Saharanpur. She has 17 years experience in academics and three years in industry. She has published around 30 research papers in international journals of repute indexed in Scopus, ESCI, Google Scholar, and other eminent databases. Currently, she is working for a funded projects, sanctioned by the Collaborative Research and Innovation Program (CRIP) under TEQIP-3 by Dr. A. P. J. Abdul, Kalam Technical University Uttar Pradesh, Lucknow. She is a member of IEEE and other professional societies such as ISTE, CSI, etc. She has been a reviewer and guest editor of several reputed journals and books. She has also served in many conferences as session chair.

Narendra Kumar has 15 years of academic experience. He has BTech (IT), MTech (CSE) and Ph.D. in Computer Science and Engineering (CSE). He has published more than 15 papers in reputed journals. He has also been an entrepreneur and has established many incubation centers. He has been the team lead for many E-learning projects. His research interests are in blockchain, AI and digital image processing. He has had leading positions in many universities and handled many international accreditations. He has established many centers of excellence in the latest technologies.

Sanjay Gour is currently working as professor and head of the Department of Computer Science and Engineering at the Jaipur Engineering College and Research Centre, Jaipur. He completed his Ph.D. in data mining and data analytics. He has published more than 90 research paper with Scopus indexed and UGC recognized journals, and has also published seven books on computer science and information management. He has successfully guided five Ph.D. candidates and five are still working under him. He organized a number of conferences/workshops as core organizer and special sessions in other countries including London and Malaysia. He is a lifetime member of the Computer Society of India and the Institution of Engineers. He has served at various universities as expert lecture of various topics including digital marketing, cyber security, data analytics etc. and also served at conferences as keynote speaker and session chair. He is recipient of the prestigious President Rover Award; Best Paper Award in IEEE sponsored international conferences; Best Paper Award in Springer sponsored international conferences; Achiever of Year Award for 2016 and 2017 by The Institution of Engineers; Appreciation Award by AICTE-Internshala-2019; Appreciation on Republic day by SDM-Kapasan Chittorgarh; Young Teacher Award Udaipur; Global Outreach Excellence Award, 2018; Best Researcher Award, 2019; National Education Icon Award, 2019; Appreciation in World Education Summit, 2020; Best Resource Person Award, 2020; and many more. He is also a national-level story writer and poet.

 **Sunil Kumar Chawla** is working as an assistant professor in the Department of Computer Science and Engineering, Chandigarh University, Mohali, Punjab, India. He is currently pursuing his doctorate from IKG Punjab Technical University. He has more than 15 years' teaching experience. His research interests lie in digital image processing, biometrics, image segmentation, and machine learning. He is a member of IEEE and other professional societies such as ISTE, CSTA, IAENG, and IETE. He has more than 30 publications in international journals of repute indexed in Scopus, ESCI, Google Scholar, and other eminent databases. He has been associated with many reputed international conferences in the capacity of Technical Programme Committee member. He has been reviewer and guest editor of several reputed journals. He has published the book *Segmentation and Normalization of Iris for Human Recognition* (Lambert Academic Publishing, Germany). He has also filed two patents and is guest editor of two books for CRC Press, Taylor & Francis Group.

1 Predicting Fraudulent Motor Vehicle Insurance Claims Using Data Mining Model

Jacob Muchuchuti and
Stewart Muchuchuti

CONTENTS

1.1 INTRODUCTION

Insurance, as a risk transfer mechanism, has become the hope for individuals, social groups, and businesses (Viaene and Dedene 2004). The benefits of insurance therefore may not be underestimated to insurance companies, policyholders, and the economy at large. However, players in the insurance industry are faced with a challenge that undermines the gains that may be realized from the use of insurance as a risk management concept as insurance fraud is on the increase hence depleting the funds paid in by the many honest customers to cover genuine losses (Insurance Europe 2013). While insurance companies have tried to develop effective procedures for identifying, investigating, and deterring fraudulent activities, according to Mosley and Kucera (2014), these initiatives can go that far. However, despite the experience that the professionals who are involved in the management of fraudulent claims may have, there are not enough of these trained eyes to review every claim, resulting in some fraudulent claims ending up slipping through the cracks, resulting in payments being made that were not supposed to be processed.

1.2 PROPOSED MODEL

The research sought to develop an input/output (black box) data mining model that aimed at assisting insurance companies to be able to predict fraudulent claims in motor vehicle insurance using observable attributes, using available algorithms.

1.2.1 STATEMENT OF THE PROBLEM

According to Chudgar and Asthana (2013), there has been an increase in fraudulent insurance cases lately and the trend has been observed from an international, continental as well as regional perspectives. The United Kingdom government estimates that the insurance industry faces about £3.4 billion of detected and undetected insurance fraud (HM Treasury 2015). In the United States of America, the Coalition against Insurance Fraud (2012) posits that insurance fraud is ranked second in the list of expensive crimes with 10 percent of the fraudulent transactions being from fraudulent claims (Insurance Information Institute 2019). Ernst and Young (2011) observed in their research that, out of the total claims that are presented in health insurance in India, nearly 25 percent are fraudulent. Africa has not been spared with the trend of fraudulent insurance claims. South Africa, being the continent's economic hub, estimates that, out of ZAR45 billion spent on claims in South Africa, about ZAR5.5 billion is lost annually through insurance fraud with 32 percent of the latter being suspected to be fraudulent claims (Risk & Insurance Zimbabwe 2018). These fraudulent claims are driving up insurance companies' overall cost resulting in their continued existence being threatened. Given the above statistics, the researchers' view is that insurance fraud is an economic *cancer* that has ravaged economies across the entire globe. We are of the view therefore that the use of predictive analytic tools such as data mining and machine learning become handy in dealing with the challenge at hand.

1.3 LITERATURE REVIEW

1.3.1 THE CONCEPT OF INSURANCE

Insurance may be considered to be as old as humanity although the way it was being used then and how it is done today may be different. There are two major types of insurance policies: life insurance that normally have a longer term to maturity that is more than a year and the short term ones that are renewed on an annual basis. Automobile insurance or motor vehicle insurance, which is the research fulcrum, belongs to the latter. Auto insurance is a contract between a policyholder (insured) and the insurance company (insurer) that protects you against financial loss in the event of an accident, or theft of your motor vehicle as well as fire (www.iii.org). The insurance company should therefore restore the insured to the same financial position he/she had before the occurrence of the loss The insurance company agrees to pay the insured's losses as outlined in your policy and auto insurance normally provides coverage for *property* in cases such as damage to or theft of the vehicle; *liability*, where an insured is legal responsible to others for bodily injury or property damage; or *medical*, being the cost of treating injuries, rehabilitation, and sometimes lost wages and funeral expenses for the dead.

1.3.2 INSURANCE FRAUD

The Insurance and Pensions Commission of Zimbabwe (IPEC) on its official web-site (www.ipec.co.zw) defines insurance fraud as cases in which individuals or entities lie to an insurance company for the sake of getting financial compensation for something they would not have received had the truth been told. The definition is supported by the Institute of Internal Auditor's International Professional Practices Framework (IPPF) as quoted by Ernst & Young (2011) that defines fraud as any illegal act characterized by deceit, concealment, or violation of trust. As mentioned earlier, insurance is based on the principle of mutual benefit and is designed to protect against significant, but uncertain losses; however, fraudulent claims undermines this system since they deplete the funds paid in by the many honest customers to cover genuine losses (Insurance Europe 2013).

Insurance fraud has devastating effects to all the insurance industry players, including and not only limited to insurance companies and their customers. Insurance fraud thus has adverse effects to the efficient functioning of a national economy depending on the magnitude of fraud. One major impact of insurance fraud to the insurance companies is that it drives up the costs of doing business and may lead to huge losses to be incurred. These costs are in the form of investigation of all the claims as well as the ultimate payment of claims that are not genuine.

It is without doubt that the insurance expense of the average household would also increase due to higher premiums that are then paid in order to cover the cost of the fraudulent transaction. This has the effect that policyholders no longer pay fair premiums and that has the resultant effect of them shunning the insurance market. This emanates from the concept of pooling where policyholders share losses.

According to Mpofu, De Beer, Nortje, and De Venter (2010), fraudulent claims have a negative impact on policyholders due to delayed processing of genuine claims, resulting in policyholders failing to get value for their money.

1.3.3 FRAUD PREDICTIVE VARIABLES

The variables, also known as attributes, are used to organize records of data in database tables. Hargreaves and Singhania (2015) in their research on fraudulent insurance grouped the predictive variables under four broad categories namely, demographic, vehicle, policy type, and claim characteristics.

Demographic characteristics according to Hargreaves and Singhania (2015) include but are not limited to the accident area (whether the accident is purported to have happened in the city or in a rural set up). The claimant's gender and the drivers' age are also critical variables including whether the claimant had a history of having changed their physical address before the fraudulent claim. Viane, Ayuso, Guillen, Gheel, and Gedene (2007) in their research conclude that information about the type and use of the insured vehicle is important as a predictor of a fraudulent claim. According to the research, characteristics include whether the motor vehicle is a van, truck, or buses. The research by Hargreaves and Singhania (2015) suggests that claims involving sedans are likely to be fraudulent. Hargreaves and Singhania (2015) also hint that the age and price of the vehicle also have an influence on fraudulent claims, suggesting that vehicles aged more than five years and valued at less than $30,000 were likely to be involved in fraudulent claims.

Research by Hargreaves and Singhania (2015) and Lincoln et al. (2003) indicate that in all fraudulent claims, the insured would have comprehensively insured the vehicle. Since an insurance fraudster is supposed to plan as mentioned earlier in this report, the fraudster would not have known the exact *modus operandi* hence comprehensively insuring the vehicle. The researchers concur with the above observations by Lincoln et al. (2003) as well as Hargreaves and Singhania (2015), the justification being that the fraudster would be leaving all possible options to present and they would not have the proximate cause to use at planning stage. It therefore makes sense to the researcher that in order to be proactive, the insureds would want to cover all perils as the fraudulent loss may end up being fire, accident or theft, as the case may be. Lincoln et al. (2003) posit that prior claims, the types of deductibles as well as when the claim is done have an influence on whether a claim is fraudulent or is genuine. This is supported by Hargreaves and Singhania (2015) who allude to the fact that before a fraudulent claim is submitted, the insured would have submitted two to four claims. As far as the researchers are concerned, prior submission of claims before a fraudulent claim would sanitize the latter as the insurance company may become complacent in processing the fraudulent claim, especially if the prior claims were considered to be non-fraudulent. As was discovered by Viane et al. (2007) the claim characteristics also include the time of the day that accidents are purported to have happened wherein in their study they noticed that that most fraudulent claims are perceived to have originated from accidents or losses that happen during weekends.

1.3.4 MACHINE LEARNING ALGORITHMS

1.3.4.1 Naive Bayes

Naïve Bayes is a powerful probabilistic supervised learning method which uses a training data set with known target classes to predict the classes of future instances (Milgo 2016). The algorithm assumes that the presence or absence of a particular variable of a data set does not depend on the presence or absence of any other attributes in the same set. This algorithm is based on Bayes Theorem of conditional probability. The Bayes theory is normally used for conditional probability when an outcome of a probability may be combined with the outcome of the other probability. In the case of the study under consideration, where there are a number of variables, a probability of a male presenting a fraudulent claim may also be conditional in the event that the male's age is within the confines of the age variables that may be a predictor. Despite its simplicity as indicated above, Muhammad (2014) alludes to the fact that the Naive Bayesian classifier often does surprisingly well and is widely used as it often outperforms a number of more sophisticated classification methods.

1.3.4.2 Decision Trees

According to Kotsiantis (2011), decision tree (DT) techniques have been widely used to build classification models because of their simplicity as well as the ability to closely resemble human reasoning. Decision tree is capable of building classification or regression models in the form of a tree structure by breaking down a data set into smaller and smaller subsets while at the same time an associated decision tree is incrementally developed. At the end of the day, the result is a tree with decision nodes and leaf nodes. With its capacity to impersonate a human being, the researchers saw sense in using DT in solving the problem at hand.

1.3.4.3 Logistic Regression

The outcome in logistic regression is measured with a variable where there are only two possible outcomes. The characteristics about logistic regression explained above makes the algorithm suitable for solving the classification problem at hand. The reasoning is that it estimates discrete values (binary values like 0/1, yes/no, true/false) based on a given set of independent variable(s). In this case the outcome is either fraudulent or non-fraudulent that may be represented by either 1 or 0 respectively. In simple terms, logistic regression predicts the probability of occurrence of an event by fitting data to a *logit function.*

1.3.4.4 Support Vector Machines (SVM)

Given a set of training examples, marked as belonging to one of either class, an SVM algorithm builds a model that predicts whether a new example falls into one class or the other, (www.cse.iitk.ac.in). This means once a machine learns the characteristics of certain variables to come up with an outcome, the machine predicts the outcome from what has been "learnt." SVM was relevant to the data set provided since the data was separated between "*training*" and "*test*" data.

1.4   METHODOLOGY

1.4.1   Data Acquisition and Description

The data set, which was retrieved from an online repository accessed on (https://databricks-prodcloudfront.cloud.databricks.com) consists of 1,000 records (claims) with 39 variables which describe the characteristics of claims that may be either fraudulent or non-fraudulent. The data set contains anonymized variables that are either numerical or categorical.

1.4.2   Data Pre-processing

Data pre-processing or data cleaning is meant to ensure that a decision is made on how to deal with missing values as well as managing categorical variables that may not be compatible with data encoding (Boodhun and Jayabalan 2018). This process involved elimination of some of the variables that do not have any predictive power as well as dealing with correlated variables. Missing values, dimensionality reduction, and feature extraction and selection were dealt with at this stage. Missing values were replaced with the most frequent values within each column. Our use of the mode method is in line with Magnami (2004), who suggests the use of mean or mode as the basis of dealing with missing values in order to reduce the influence of exceptional data.

The following Python code snippet shows that the missing values that were represented by 'n/a', 'na', '---', or '?' were considered as missing and were replaced with the most frequent values in the particular column.

The number of variables was reduced in order to increase the efficiency of the model. The process, known as dimensionality reduction involved feature extraction which is applied to transform the high dimensional data into fewer dimensions to be used in building the models and feature selection which involves the selection of prominent variables (Boodhun and Jayabalan, 2018).

To that end, a two-pronged approach was used to eliminate irrelevant and redundant variables. The first was based on the researchers' understanding of the relevancy of the data based on their exposure to the insurance industry. These variables were eliminated because their significance was deemed to be low in the modeling process. The second set of variables was eliminated based on the level of multicollinearity or correlation coefficient between individual variables. This was in line with the Taber (2018)'s suggestion that if a correlation coefficient between two variables is higher than 0.6 or 0.7, one of them could be dropped from the data set to reduce its dimension and to reduce redundancy (www.uta.fi). As a result, variables such as property claim, injury claim, vehicle claim, and property claim amounts were dropped from the data set in order to hence enhance the performance of the model.

1.4.3   Encoding

The data was presented into a language that could be understood by the computer, using a process called encoding. This involved the transformation of categorical

---

#Loading dataset and detecting missing values. n/a, na,?, detected as missing values
missing_values = ["n/a," "na,"" ?," "--"]
df = pd.read_csv("/Users//jacobm//Desktop//Datasets//fraud.csv," na_values=missing_values)

FIGURE 1.1 Reduction in variables.

variables to binary or numerical counterparts. This notion is supported by Ferreira (2018) who suggests the need to ensure that most input variables be converted to numerical in order for many machine learning algorithms to recognize them. One – hot or label encoding – was used. Predictive variables (such as type of car) that may have more than one value were transformed into numerical values by using numerical values to represent them, for example, a Mercedes could be represented by 1, Toyota represented by 2. Label encoding, an important pre-processing step for the structured dataset in supervised learning, refers to the conversion of data labels into numeric form so as to convert it into machine-readable form (www.geeksforgeeks.org). The researchers used label encoding to convert the target variable from categorical to numerical. With supervised learning, the target variable or outcome is already known (whether claim is fraudulent or not) and the idea hence is to predict whether a particular claim is fraudulent or legitimate using the attributes that the machine would have been trained to observe. In this case then, the outcome of "fraudulent" or "legitimate" was encoded into numerical values where fraudulent is represented by 1 and legitimate is represented by 0.

1.5 APPLICATION OF CLASSIFICATION ALGORITHMS

Four algorithms were used to train the data using a data ratio of 70:30 being training and testing respectively. It therefsssore means that 700 records (claims) were used to train the model and 300 were used to check the accuracy of the algorithms. The fact that the outcome (whether fraud or not) was known from the data set, the machine was trained to understand the attributes of a fraudulent claims as well as the genuine claim, given the attributes of each of the two outcomes. The training was done on all the four algorithms, Decision Trees, Naïve Bayes, Logistic Regression, and Support Vector Machine, and the performance of each algorithm was tested for accuracy. The latter involved taking the testing dataset and running it on the algorithm and assessing the extent to which it could correctly predict whether a claim was fraudulent or not, given the already known outcome.

1.6 RESULTS

1.6.1 ATTRIBUTES OF PREDICTOR VARIABLES

Attributes such as demographic, policy type, vehicle type, and the claim characteristics were observed in claims that were predicted as fraudulent. On demographic characteristics, it was noted that policyholder's hobbies such as camping and chess were considered to be prone to perpetrate insurance fraud. This is in addition to such attributes such as the claimant's age, gender, educational level, occupation that were suggested by Hargreaves and Singhania (2015).

On the policy type characteristics, the current study suggests that vehicles that were only insured for vehicle theft, single car collision and multi vehicle collision as well as parked car were victims of fraudulent claims. These results are a diversion from the studies that were done by Lincoln et al. (2003) and Hargreaves and Singhania (2015) who hinted that fraudulent claims were observed in comprehensively insured vehicles.

On the claim characteristics, attributes such as the amount claimed, property and body injuries claims as well as the severity of the incident were confirmed to be some of the attributes that may be observed in fraudulent insurance claims. These results seem to be an addition to assertions by Lincoln et al. (2003) who suggest that for a claim to be suspicious of being fraudulent, the claimant is supposed to have made prior claims in the past.

1.6.2 Variables with the Most Predictive Influence

After running the model based on six best predictive variables, characteristics such as the insured's hobbies and incident severity were noted as having the most predictive power. Claimants whose hobbies include camping, chess, and cross fit were considered to be prime suspects for fraudulent claims. The researchers concur with the results of the analysis considering that people who are into camping may be considered to have high risk appetite because of the exploratory nature of the hobby, the same can also be said for chess, a game that requires strategy and requires one to be calculative. This is supported by research by Levi (2008b) that suggested that specialist skill is a requirement for a person to perpetrate insurance fraud. In addition, the research brought a dimension that whether the damage incurred is trivial, minor, or large, the claim is potentially fraudulent. The following Python snippet is evidence of the most influential predictor variables as per analysis.

1.6.3 The Classification Algorithms Used

The four most popular algorithms, that is, Decision Tree, Support Vector Machine, Naïve Bayes and Logistic Regression, were used in the training of the model and each of them was tested for predictive efficiency. Table 1.1 shows the efficiency rates for each algorithm.

Creating a new Dataframe based on 6 best features selected by Recurssive
Feature elimination
*X_rfe = X.filter(["insured_hobbies_camping," "insured_ hobbies_chess," "insured_hobbies_*
cross_fit,"
"Incident_severity_Major Damage," "incident_severity_Minor Damage,"
"incident_severity_Trivial Damage"])

FIGURE 1.2   Creating new data frame.

TABLE 1.1
Algorithms predictive efficiency rates

Algorithm	Predictive efficiency rate (%)
Decision Tree	83.3
Support Vector Machine	76.7
Naïve Bayes	69.3
Logistic Regression	76.3

Source: Research results.

The Decision Tree algorithm was therefore used to develop the model based on the level of predictive accuracy and the model may be presented mathematically as follows:

$$\text{FIC}_{MV} \approx F \text{ \{Demographic attributes; Policy type attributes;}$$
$$\text{Claim characteristics; Vehicle type characteristics\}}$$

Where FIC_{MV} = Fraudulent Insurance Claim for Motor Vehicle Policy

1.7 CONCLUSIONS AND FUTURE SCOPE

Having synthesized the results and subsequent discussion, the following conclusions were drawn from the study:

i)   Males and females alike are capable of committing insurance fraud with statistics showing that the latter are more prone to committing the offence as compared to the former.
ii) Policyholders between the age of 30 and 50 years are the main culprits as far as making fraudulent claims.
iii) Policyholders from across the academic and professional qualifications are all bound to be perpetrators of insurance fraud by submitting fraudulent insurance claims.
iv) Policyholders' hobbies and the extent of the damage on the vehicle have the most predictive influence.

1.7.1 RECOMMENDATION OF FUTURE STUDIES

Further research may be done to establish the results of the model are consistent if other classification algorithms are used in developing the model. These include the genetic algorithms and Artificial Neural Network. Deep learning algorithms such as Deep Neural Networks may also be considered for further study to establish if the results of the model improve. Furthermore, consideration may be made to come up to come up with a model that seeks to predict whether a policyholder is a potential insurance fraudster or not based on given attributes. This would assist insurance companies to come up with interventions for all such customers that are considered to be '*potential fraudsters*' right at the underwriting stage unlike waiting to identify such traits at the claiming stage. The idea is to ensure that interventions are instituted from the point the risk is accepted so that decisions may be done on whether to accept the risk with conditions, accept as is or reject the proposal based on the observable demographics of the proposer.

REFERENCES

Boodhum and Jayabalan, 2018. Risk Prediction in Life Insurance Industry Using Supervised Learning Algorithms. Unpublished article.

Chudgar and Asthana, 2013. Life Insurance Fraud – Risk Management and Fraud Prevention. *International Journal of Marketing, Financial Services and Management Research ISSN 2277-3622*, 2(5).

Coalition against Insurance Fraud, 2012. Go Figure, Go Data (accessed on www.insurancefraud.org/stats.htm),

Ernst & Young, 2011. *Fraud in Insurance on the Rise Survey 2010–11*.

Ferreira, 2018. How to easily implement One-Hot Encoding in Python. Unpublished.

Gepp, A., Wilson, J. H., Kumar, K., and Bhattacharya, S. (2012). A comparative analysis of decision trees vis-à-vis other computational data mining techniques in automotive insurance fraud detection. *Journal of Data Science*, 10(3), 537-561 Accessed on www.jds-online.com/volume-10-number-3-july-2012

Hargreaves and Singhania, 2015. Analytics for Insurance Fraud Detection: An Empirical Study. *American Journal of Mobile Systems, Applications and Services*, 1(3): 227–37.

HM Treasury, 2015. Insurance Taskforce Report (accessed on www.gov.uk).

Insurance Crime Bureau, 2017.

Insurance Europe, 2013. The Impact of Insurance Fraud.

Kotsiantis, 2011. Decision Trees: A Recent Overview. Unpublished.

Levi, 2008b. *The Phantom Capitalists: The Organisation and Control of Long Firm Fraud.* Revised Edition. Unpublished.

Lincolin, Wells, and Petherick, 2003. *An Exploration of Automobile Insurance Fraud.* Bond University.

Maghami, 2004 *Techniques for Dealing with Missing Data in Knowledge Discovery Tasks*, Department of Computer Science, University of Bologna.

Mosley and Kucera, 2014. Analytics for Claim Fraud Detection. Unpublished.

Mpofu, De Beer, Nortje, and De Venter (2010). *Investment Management*. Third Edition, Van Schaik Publishers, Pretoria, South Africa

Muhammad, 2014. Fraud: The Affinity of Classification Techniques to Insurance FRAUD Detection. *International Journal of Innovative Technology and Exploring Engineering, Issue 10 Vol 3 ISSSN 2278-3075*

Risk and Insurance Zimbabwe, 2018. Fraudulent Claims: Insurers lose 30% of revenue. www.ir.co.zw (accessed 7 December 2018).

Taber, K. 2018. The Use of Cronbach's Alpha When Developing and Reporting Research Instruments in Science Education. *Research in Science Education*, 48: 1273–96.

Viane, Ayuso, Guillen and Gedene, 2007. Strategies for Detecting Fraudulent Claims in the Automotive Insurance Industry. *European Journal of Operational Research*, 176(1): 565–83.

Viane, Derrig, and Dedene, 2004. A Case Study of Applying Boosting Naïve Bayes to Claim Fraud Diagnosis, *IEEE Transactions on Knowledge and Data Engineering*, 16(5).

www.cse.iitk.ac.in accessed on June 16, 2019.
www.iii.org accessed May 15, 2019
www.ipec.co.zw accessed January 5, 2018.
www.kaggle.com accessed on June 16, 2019.
www.saedsayad.com accessed June 24, 2019.

2 Novel 8: 1 Multiplexer for Low Power and Area Efficient Design in Industry 4.0

Tripti Dua, Anju Rajput, and Sanjay Gour

CONTENTS

2.1 INTRODUCTION

The scenario around the world has been on the daily basis based on the modification and requirement of human population, this leads to next industrial revolution. Nowadays, every industry in engineering is transforming into digitized form. In addition, manufacturing of products is being done for mass production (Frank and Dalenogare 2019). Earlier, Industry 1.0 was introduced in the eighteenth century, in which development was made in sectors like energy and transportation with the help of steam engines and steamboats. Industry 2.0 was introduced in the nineteenth century in which mass production began (Oztemel and Gursev 2020). Also, electricity was

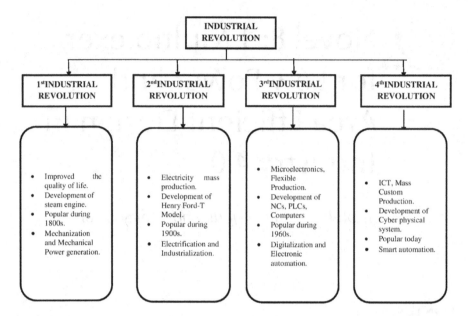

FIGURE 2.1   Industrial revolution.

emerged which lead to the development of industries like automotive, oil, and agriculture. Industry 3.0 was introduced in the 2000s where computer and Internet came into the wide picture. Development in technology has not stopped since back then, which lead to introduction of Industry 4.0 (Raut, Gotmare, Narkhede, Govindarajan, and Bokade 2020). Different phases of industrial revolution are demonstrated in Figure 2.1.

Recently, advancement in technology is being made at a very high pace in which Industry 4.0 plays a very vital role. The cause of the Industry 4.0 revolution is an increase in digitization, interconnectivity between different products, supply chains, and different business models. Industry 4.0 provides communication between the physical and digital world. Industry 4.0 means a combination of two different worlds, i.e., real and virtual; it means amalgamation of conventional and modern techniques in manufacturing (Rojko 2017). Industry 4.0 is not only engineering but also creativity, management, science, knowledge, and application. Two things are needed for Industry 4.0, i.e., operation's environment and Industry 4.0 ready engineers. Pillars of Industry 4.0 are augmented reality (AR) or virtual reality, additive manufacturing with the help of 3D printing, Internet of Things (IoT), big data analysis, cloud computing, advanced simulation, autonomous robots by artificial intelligence (AI), and cyber security system. With the help of AR digital data as well as perceptual information can be displayed by some devices, which help in numerous fields. 3D printing provides flexibility in designing complex structures. It is used to produce prototypes as well as products, which are needed in less quantity. This leads to reduction in costs and production time (Aceto, Persico, and Pescape 2019). The Internet of Things (IoT) provides communication between different machines by accessing and transmitting their data. This leads to faster decision making and helps in making new business models. Big data analysis (Qi and Tao 2018) is cleaning, organizing, visualizing, analyzing, and optimizing massive data that is available in industry in an efficient way. Cloud computing helps in retrieving, storing, and transmitting data. It also enables to share, manage, and compute data with robustness. With the help of artificial intelligence (AI), robots can be designed which work autonomously and have abilities like humans and manage and transmit data. This reduces the need for manpower. Cyber security provides protection to information and any industrial system from any cyber breach such as piracy and corrupting of hardware software or data lines of any system (Ayaz, Ammad-Uddin, Sharif, Mansour, and Aggoune 2019).

The complete process of producing a new product in the market can be understood by a flow chart (see Figure 2.2). Technology is getting advance in terms of speed, required area, and power consumption (Zambon, Cecchini, Egidi, Saporito, and Colantoni 2019).

Some of the advantages of implementing Industry 4.0 can be listed as increment in profits, time reduction in introducing novel products to the world, reduction in costs due to effectiveness and efficiency and reduction in occurrence risk (Aceto, Persico, and Pescape 2019). Figure 2.3 indicates the functioning of a smart factory using all the pillars of Industry 4.0.

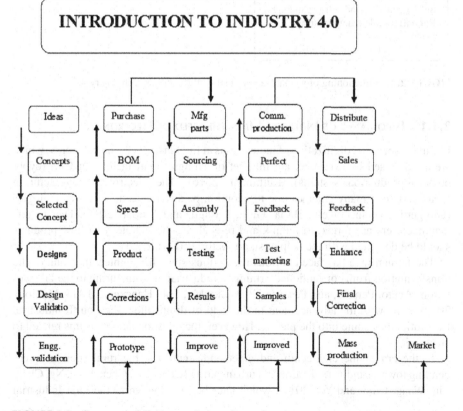

FIGURE 2.2 Process to deliver a product to the market.

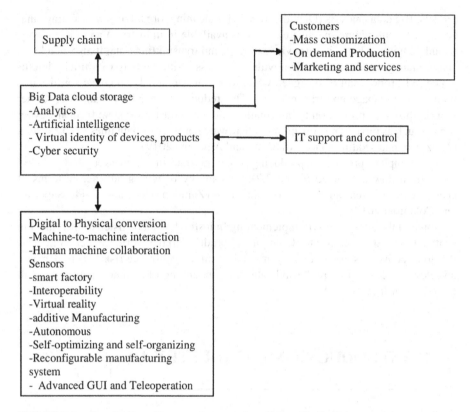

FIGURE 2.3 Functioning of a smart factory using all the pillars of Industry 4.0.

2.1.1 IMPORTANCE OF INDUSTRY 4.0 IN THE FIELD OF ELECTRONICS

In current scenario, digitization of industries is at full blast. The basic keywords, which are almost used by everyone, are Internet of Things, smart factory, or CPS (cyber physical/production systems) (Rüttimann 2016). The electronics-manufacturing sector has seen many changes due to adoption of different innovative technologies (Giorgetti. Lucchi, Tavelli, Barla, Gigli, Casagli, and Dardari 2016). Due to this manufacturers are forced to rethink and re-evaluate their strategy of how products should be designed with the help of smart factories (Figure 2.4).

The human race has faced various revolutions since the eighteenth century. The transformation from conventional farming to industrial manufacturing to the IT revolution (Aceto, Persico, and Pescape 2019; Liu, Ma, Shu, Hancke, and Abu-Mahfouz 2020). Now with the advancement in the field of digitization, the Fourth Industrial Revolution has come into the picture. However, the question is: how is this related to electronics?

In the era of the eighteenth and nineteenth centuries, the rural societies were bending towards industrialization and urbanization (Chen, Lin, Chen, Liao, Ng, Chan, Liu, Wang, Chiu, and Yen 2019), which later became known as the First Industrial

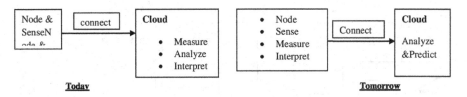

FIGURE 2.4 Comparison between today's and tomorrow signal chain.

Revolution. The focus of this revolution was on development of water and steam power engine (Kovacs and Husti 2018). After that, the Second Industrial Revolution (1870–1914) came into existence, which mainly focused on mass production using electrical energy. The major advancement during this era included telephones, light bulbs, phonographs and IC engines (Zambon, Cecchini, Egidi, Saporito, and Colantoni 2019).

Then at the beginning of 1980 the Third Industrial Revolution brought the advancements in the field of usage of PLCs, personal computers, Internet, and information and communications technology. This revolution was followed by Industry 4.0, which links societies with the human body. The base of Industry 4.0 is advancement in communication and connectivity within and between societies. This new age brings many prominent breakthrough technologies, due to which human life is becoming easier. It includes robotics, artificial intelligence, nanotechnology, renewable energy, quantum dots, 5G networks, 3D printing, Internet of Things, decentralized consensus, and automization of vehicles. Semiconductor industries use this concept to modify their products with the help of automated smart factories. The current trend is also based on optimization. These optimization techniques adopted by industries, via automation in various sectors, have transformed the whole communication system and also result in increasing productivity with minimal human resource (Miranda, Ponce, Molina, and Wright 2019).

More flexibility occurred when the AI printed circuit board came into existence, which further created connected objects, and smart home equipment. This means the new printed circuit board is also incorporated within the features of artificial intelligence. Now, the Internet of Things has succeeded in connecting every machine for transmission of data among the different sectors present in a company. Using this technology new business models can be created. The electronics industry uses this data to build relation with the customers, which helps them to build devices for their customers (Xu, Xu, and Li 2018).

Micro-electromechanical system (MEMS) also came into existence with the development of miniature products like smartphones. After that, cloud computing and IoT came into the picture (Ayaz, Ammad-Uddin, Sharif, Mansour, and Aggoune 2019). Nowadays, MEMS sensors are used in automation and manufacturing in the fields of defense, robotics, oil and gas, energy, and agriculture giving rise to an industrial revolution with the help of Industry 4.0 (Javaid, Haleem, Vaishya, Bahl, Suman, and Vaish 2020). A conventional MEMS based smart sensor is shown in Figure 2.5.

We are in the midst of digitization of manufacturing. Industry 4.0 is the fourth revolution, which occurred in the field of manufacturing (Rubmann, Lorenz, Gerber,

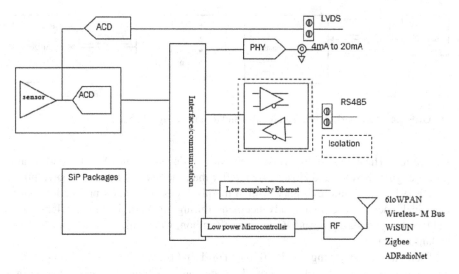

FIGURE 2.5   MEMS-based smart-sensor-solution.

Waldner, Justus, and Engel 2015). This fourth industrial revolution is accommodating smart and autonomous systems fueled by data and machine learning. Industry 4.0 is the optimization of computerization, which was introduced in Industry 3.0. Now apart from computerization, decision making without human involvement is more focused criteria. Now the focus is more on developing smart machines, which can access more data but without consuming large space (Ahmed, De, and Hussain 2018). Therefore, there is a need to develop area-efficient machines so that our industry will become more efficient and productive and less wasteful. As Industry 4.0 is digital transformation of manufacturing and processes, multiplexer design with relatively less number of transistors results in miniature products.

In the digital world, a connection between hardware and software is required so that optimization of development of a novel product can be obtained. Multiplexer is one of the basic building blocks in any digital circuit, which provides parallel to serial data conversion in communication of signals in most of the circuits designed in pillars of Industry 4.0 like artificial intelligence and IoT (Elijah, Rahman, Orikumhi, Leow, and Hindia 2018).

Multiplexer is a combinational circuit, which receives parallel binary information and transfers it through a single line. Multiplexer can be used in optical communication based Artificial Immune System (AIS) which has broad applications in economy analysis, environmental techniques, and improvements in processing.

2.1.2   APPLICATIONS OF MULTIPLEXER

Memory

Multiplexers are utilized as memory to store a large amount of data in many devices such as computers, smartphones, robots etc. It assists in reducing the number of copper lines required to connect memory and other part of the device.

Communication System

For transmission of different signals such as audio, video signals simultaneously through a single channel, a multiplexer is implemented in the system. This leads to an increase in efficiency of any system.

Telephone Network

For the transmission of multiple audio signals over a single line in telephone network, a multiplexer is used. All audio signals which are transmitted through a multiplexer are separated and the desired signal is received by the receiver.

Satellite Transmission

The transmission of signals from spacecraft or any other satellite to earth equipment is carried out with the aid of multiplexers.

2.2 INTRODUCTION TO BASIC MULTIPLEXER

The multiplexer is a digital switch that takes 2^n inputs, uses 'n' select lines and produces a single output. In conformity with the binary combination of the select lines, output is produced by transmitting the selected data input line to the output line. 2:1 MUX is the simplest multiplexer (Metzgen, and Nancekievill 2005) which can be used to design higher order multiplexer. Any multiplexer can be designed with different design styles at transistor level, i.e., by Transmission Gate Logic (TGL), Pass Transistor Logic (PTL) or CMOS logic. In the next section, 2:1 MUX is devised using all these three design styles and then they are compared after simulation on HSPICE tool at 1V power supply in terms of transistor count and power consumption at different technologies (Suzuki, Kawano, Nakasha, Yamaura, Takahashi, Makiyama, and Hirose 2007).

2:1 MUX takes two input data lines, has a single select line and a single output line as well. When the select line is applied with logic, '0' then input D0 is connected to the output line. When the select line is applied with logic '1', then data line D1 is connected to the output line and transmitted through it (Sarita and Hooda 2013). The Boolean expression, block diagram and logic diagram of 2:1 MUX are shown below in Figures 2.6 and 2.7 respectively.

$$y = D_0\overline{S} + D_1S$$

Truth table of 2:1 MUX with all possible combinations of input lines and select line is shown in Table 2.1. It can be deduced from the table that whenever the select line is at logic '0', then output is replica of as data input 'D0' disregarding the values of other data input. Similarly, notwithstanding the value of data input line D0, when

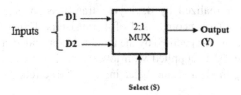

FIGURE 2.6   Block diagram of 2:1 MUX.

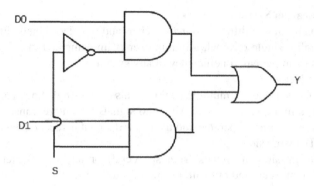

FIGURE 2.7   Logic diagram of 2:1 MUX [21].

TABLE 2.1
Truth table of 2:1 MUX

S	D0	D1	Y
0	0	0	0
0	0	1	0
0	1	0	1
0	1	1	1
1	0	0	0
1	0	1	1
1	1	0	0
1	1	1	1

the select line is provided with logic '1', then replica of data input D1 is acquired at the output line.

2.2.1   TRANSMISSION GATE LOGIC BASED 2:1 MUX

A transmission gate is a symbiotic switch that is fabricated using a PMOS transistor and an NMOS transistor. It halts or carries a signal from the input line to the output line depending on the control signal employed extraneously. The control signal provided to the gates of the transistors in a fashion such that their logical values are complementary to each other. Hence, either both the transistors are active mode or both of them are in cut-off mode concomitantly.

A 2:1 multiplexer realized by deploying transmission gate logic is shown in Figure 2.8. The circuit is encompassed of 4 transistors. Either data line A or B is passed to the output line as specified by the logic value provided at the control line 'S'.

When select line 'S' is supplied with low logic, then transistors N1 and P1 are driven in conducting mode and start behaving like short circuits whilst transistors N2

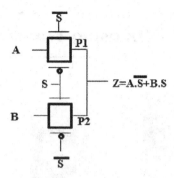

FIGURE 2.8 2:1 MUX using TGL [30].

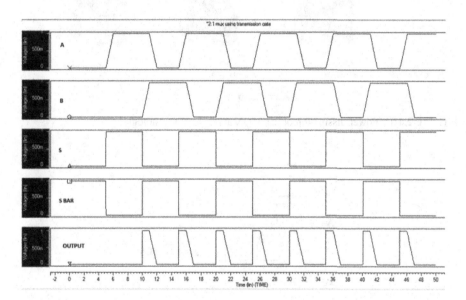

FIGURE 2.9 Waveforms of 2:1 MUX using TGL.

and P2 are driven into cut-off mode and start behaving like open circuits. Therefore, input data A is progressed to the output. When select line 'S' is supplied with high logic, then transistors N1 and P1 are triggered off and start behaving like open circuits and transistors N2 and P2 are triggered off and start behaving like short circuits. Thus, only conducting path is accessible connecting data input B and the output line. Simulation of the circuit is carried out on HSPICE tool at 45nm, 32nm, and 16nm technologies by providing 1V power supply and expected output is acquired. Waveforms attained after simulation are manifested in Figure 2.9.

Power dissipated by 2:1 MUX implemented by TGL at 45nm, 32nm, and 16nm technologies is evinced in the form of bar graph shown in Figure 2.10.

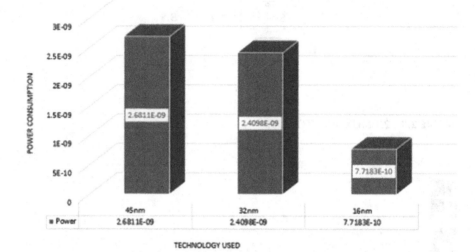

FIGURE 2.10   Power dissipation of 2:1 MUX at different technologies.

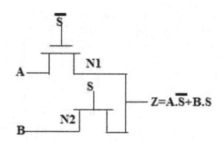

FIGURE 2.11   2:1 MUX using pass transistor logic.

2.2.2   2:1 MUX Using Pass Transistor Logic

As the name indicates, Pass Transistor Logic (PTL) utilizes NMOS transistors to transfer or occlude the electrical signal in conformity with the signal supplied at select line provided at the gate of the transistor. Number of transistors used to design any logic circuit with the help of PTL is reduced up to a great extent by eliminating redundant transistors, but this costs the performance of the circuit (Mishra and Akashe 2013). With the assistance of Pass Transistor logic, a 2:1 multiplexer can be fabricated with the aid of just two NMOS transistors. The circuit designed is shown in Figure 2.11

Pass Transistor N1 and N2 are provided with data inputs A and B respectively. When control signal S is at logic 0, then transistor N1 is activated and behaves like

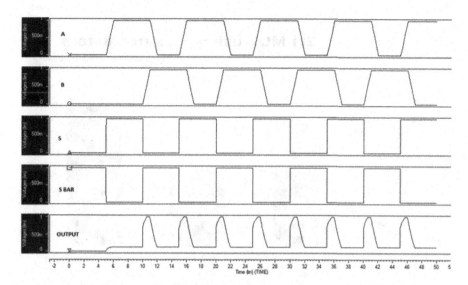

FIGURE 2.12 Waveforms of 2:1 MUX using pass transistor logic.

a short circuit whilst transistor N2 is triggered off and portrays like an open circuit. Therefore, data input A is passed to the output line. Although, when control signal S is at logic 1, then transistor N1 is triggered off whereas transistor N2 gets activated and behaves like short circuit. Thus, a conducting path is obtained only linking data input B and output line and therefore input B gets followed by output. Input and output waveforms retrieved after simulation of the circuit at 1V power supply is exhibited in Figure 2.12.

Simulation of circuit is carried out at 45nm, 32nm, and 16 nm technologies and the comparison in terms of power dissipation by the circuit at different technologies is represented by a bar graph (Figure 2.13)

2.2.3 2:1 MUX Using CMOS Logic

Complementary Metal Oxide Semiconductor (CMOS) logic deploys uniform number of both categories of MOSFETs, i.e., PMOS and NMOS. This helps in gaining better performance of any logic circuit since NMOS is pull down device and PMOS is pull up '1' device (Gupta, Arora, and Singh 2012). Therefore with zero distortion, CMOS gives complete '1' and complete '0' logics at the output. 2:1 MUX is implemented using CMOS logic with the aid of 10 transistors as shown in Figure 2.14.

Simulation is carried out at 45nm, 32nm, and 16 nm technologies with the succor of HSPICE tool and the waveforms acquired for inputs and outputs are presented below in Figure 2.15.

A comparison is made in terms of power consumption by the circuit at those technologies, which is shown with the help of bar graph representation (Figure 2.16) in the succeeding section. Thus, it can be concluded that as technology is reduced, power consumption is also reduced.

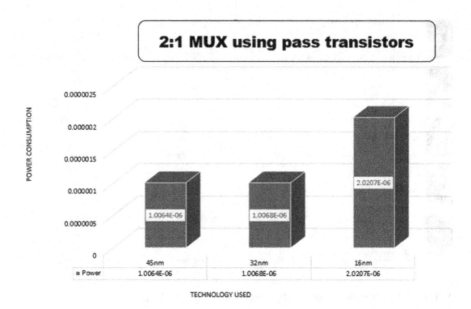

FIGURE 2.13   Power dissipation of 2:1 MUX at different technologies.

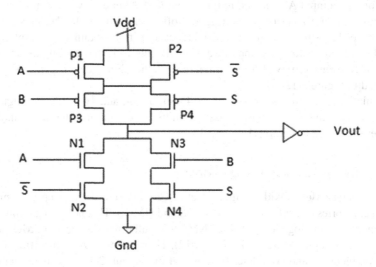

FIGURE 2.14   2:1 MUX using CMOS logic.

2.2.4   COMPARATIVE ANALYSIS

The implementation of 2:1 MUX is carried out using Transmission Gate Logic (TGL), Pass Transistor Logic (PTL), and CMOS logic. Simulation of all the circuits are accomplished by operating them on HSPICE tool at 45nm, 32nm, and 16nm technologies with 1V power supply and power dissipation in each case is monitored. Power consumed by multiplexer designed by making use of a transmission gate

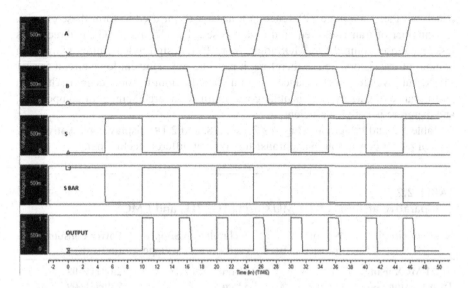

FIGURE 2.15   Waveforms of 2:1 MUX using CMOS logic.

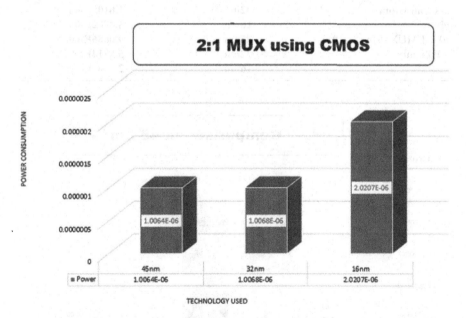

FIGURE 2.16   Power dissipation of 2:1 MUX at different technologies.

as well as CMOS logic is exceptionally low as compared to that designed by Pass Transistor Logic.

After making comparison amidst all the three circuits with different design styles, the minimum power is dissipated by 2:1 multiplexer employing TGL. However, power consumed by the circuits created by utilizing TGL and CMOS is approximately equal, but there is a huge difference in transistor count used to form these circuits. The number of transistors required to compose 2:1 MUX using TGL is three times less than those required by CMOS logic. Since PTL design style requires a minimum number of transistors, i.e., 2; therefore, it provides area efficient logic circuit for 2:1 MUX, but gives low performance due to distorted output to some degree. Although performance of CMOS logic is quite good, it utilizes a large number of components, and therefore is not area efficient.

Table 2.2 and bar graphs (Figures 2.17, 2.18, and 2.19) display comparative analysis of power consumption and transistor count at different technologies

TABLE 2.2
Comparative analysis of 2:1 MUX using TG, PTL, and CMOS

Name of circuit	No. of transistors used	Technology used	Supply voltage	Power consumption (in watts)
(i) 2:1 MUX using Transmission Gate only	4	45nm	1v	2.6811E-09
		32nm		2.4098E-09
		16nm		7.7183E-10
(ii) 2:1 MUX using pass transistors only	2	45nm	1v	1.0064E-06
		32nm		1.0105E-06
		16nm		8.4802E-06
(iii) 2:1 MUX using CMOS only	10	45nm	1v	9.0686E-09
		32nm		5.8714E-09
		16nm		2.3305E-09

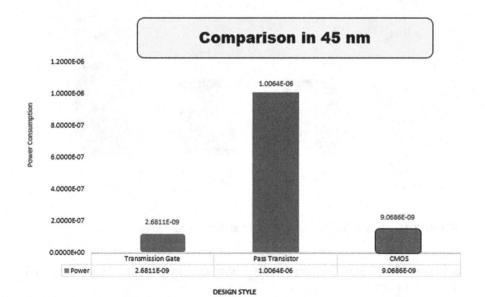

FIGURE 2.17 Comparison of power dissipation of 2:1 MUX at 45nm technology.

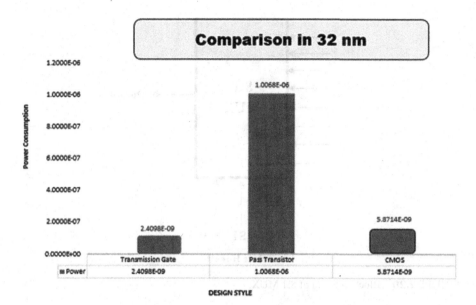

FIGURE 2.18 Comparison of power dissipation of 2:1 MUX at 32nm technology.

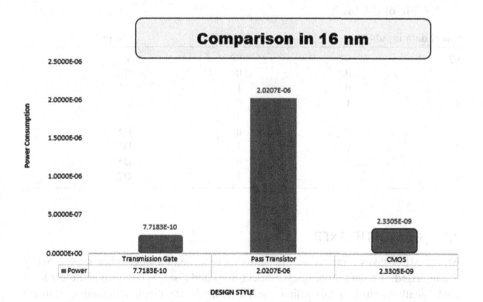

FIGURE 2.19 Comparison of power dissipation of 2:1 MUX at 16nm technology.

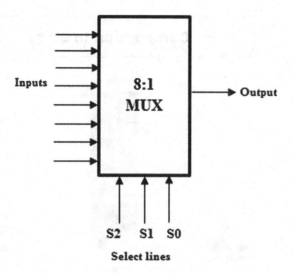

FIGURE 2.20   Block diagram of 8:1 MUX.

TABLE 2.3
Truth table of 8:1 MUX

Select data inputs			Output
S2	S1	S0	Y
0	0	0	D0
0	0	1	D1
0	1	0	D2
0	1	1	D3
1	0	0	D4
1	0	1	D5
1	1	0	D6
1	1	1	D7

2.3   8:1 MULTIPLEXER

An 8:1 MUX takes 8 inputs and transmits them through a single channel. 8:1 MUX is comprised of 8 data input lines, 3 select lines and a single output line. Select lines are basically control signals which select a particular data input and connect it to the output line (Roohi, Khademolhosseinei, Sayedsalehi, and Navi 2011). The block diagram of 8:1 MUX is shown in Figure 2.20.

It can be visualized from the truth table of 8:1 MUX (Table 2.3) that at each combination of select lines, corresponding data line is activated and is linked to the output line to provide replica of that data input to the output.

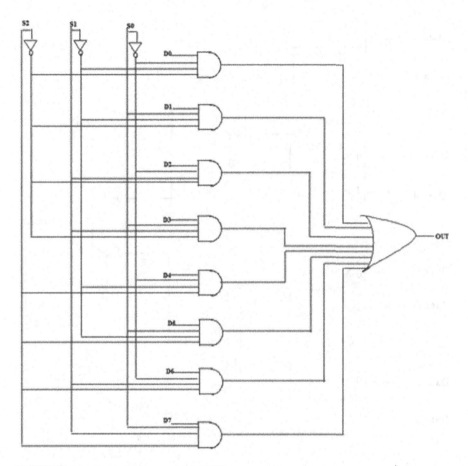

FIGURE 2.21   Logic Diagram of 8:1 MUX.

With the aid of the truth table, Boolean expression for 8:1 MUX can be deduced as follows and its logic diagram can be designed with the help of 8 AND gates, 1 OR gate and 6 NOT gates which is displayed in following section (Figure 2.21).

$$Y = D0.\overline{S2}.\overline{S1}.\overline{S1} + D1.\overline{S2}.\overline{S1}.S0 + D2.\overline{S2}.S1.\overline{S0} + D3.\overline{S2}.S1.S0 + D4.S2.\overline{S1}.\overline{S0}$$
$$+ D5.S2.\overline{S1}.S0 + D6.S2.S1.\overline{S0} + D7.S2.S1.S0$$

2.3.1   8:1 MULTIPLEXER USING LOWER ORDER MULTIPLEXERS

Since it is established that lower order multiplexers can be utilized to construct higher order multiplexers. Thus, 8:1 MUX can be implemented using seven 2:1 MUX in the manner depicted below. In the following design displayed in Figure 2.22, data 1 to data 8 are eight data input lines and A, B, and C are select lines.

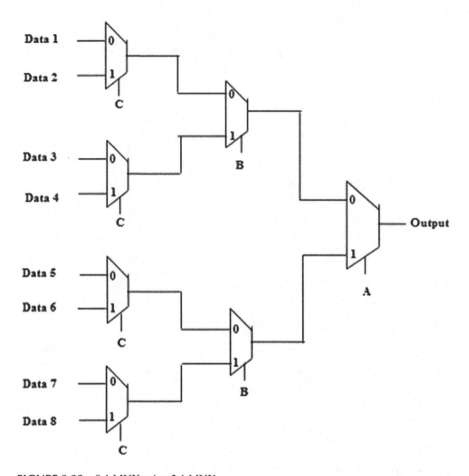

FIGURE 2.22   8:1 MUX using 2:1 MUX.

The logic diagram is comprised of three stages. At first stage four 2:1 multiplexers are used, at second stage two 2:1 multiplexers are used and at third stage one 2:1 multiplexer is utilized. At the first stage same select line C is applied to all the four multiplexers, at the second stage the same select line B is applied to both the multiplexers and at the last stage select line A is applied.

CASE I: When A=0, then 2x1 Multiplexer at third stage produces output generated by upper multiplexer employed at second stage. If B=0, then one of the 2 inputs data1 or data2 will be selected on the basis of logic provided at select line C and is passed to the output, else one of the two inputs data3 or data4 is connected to output based on the logic applied at select line C.

CASE II: When A=1, then 2x1 Multiplexer at third stage gives rise to the output generated by lower multiplexer employed at second stage. If B=0, then one of the 2 inputs data5 or data6 will be selected on the basis of logic provided at select line C and is passed to the output, else one of the two inputs data7 or data8 is transmitted to the output line based on the logic applied at select line C.

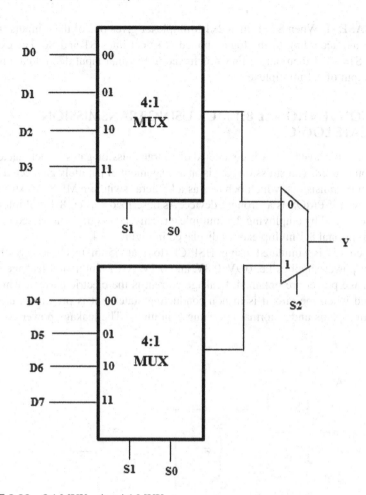

FIGURE 2.23   8:1 MUX using 4:1 MUX.

Similarly, 8:1 MUX can also be implemented by making use of combination of two 4:1 MUX and a 2:1 MUX in the following manner [27]. In the diagram shown in Figure 2.23 d0 to d7 are input data lines, S0, S1, and S2 are select lines and Y is output line.

Select lines S1 and S0 are applied to both 4:1 Multiplexers. Upper 4:1 Multiplexer takes inputs d0 to d3 and lower multiplexer takes inputs d4 to d7. Consequently, each 4:1 Multiplexer provides an output in accordance with the logic applied at the select lines.

At the first stage, the outputs of both the multiplexers act as inputs of 2x1 Multiplexer, which is available in the succeeding stage. The third select line S2 is provided to 2x1 Multiplexer.

- **CASE I:** When S2=0, then 2x1 Multiplexer produces one of the 4 inputs from d3 to d0, depending upon the logic applied at select lines S1 and S0. For example, if S1=S0=0 then output line Y is connected to the data input d0 and pass it to the output of 2:1 multiplexer.

- **CASE II**: When S2=1, then 2x1 Multiplexer gives one of the 4 inputs from d7 to d4, according to the logic applied at select lines S1 and S0. For example, if S1=S0=1 then output line Y is linked to the data input d7 and pass it to the output of 2:1 multiplexer.

2.4 CONVENTIONAL 8:1 MUX USING TRANSMISSION GATE LOGIC

The design displayed below is concocted of 14 transmission gates, which makes transistor count as 28. Transmission gate is an arrangement of parallelly connected PMOS and NMOS transistors, which behaves as a bilateral switch. NMOS is a strong zero device, even if PMOS is a strong 1 device. As described earlier, 8:1 multiplexer can be implemented by employing 2:1 multiplexer. This technique is harnessed to design the conventional 8:1 multiplexer as displayed in Figure 2.24.

The circuit is simulated using HSPICE tool at 45nm technology with three different power supplies i.e. 0.9V, 0.8V, and 0.7V. After simulation leakage current and leakage power are obtained. Leakage current is the electric current, which gets generated when the circuit is in non-conducting state or it is present in unwanted conducting paths under normal operating conditions. The leakage power comprises

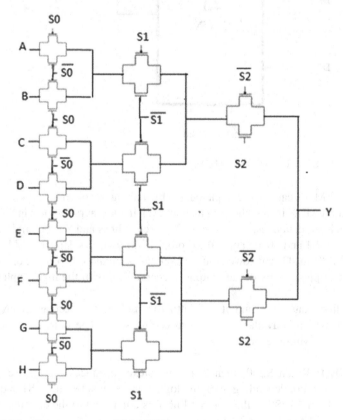

FIGURE 2.24 Existing circuit of 8:1 MUX.

of static, dynamic power dissipation of a circuit when it is in standby mode only. It can be calculated by the given expression in which I leak is the leakage current produced by the circuit and VDD is the power supply provided to the circuit (Dixit, Khandelwal, and Akashe 2014).

$$Pleak = I\ leak \times VDD$$

Input output waveform generated after simulation for conventional 8:1 MUX is exhibited in Figure 2.25.

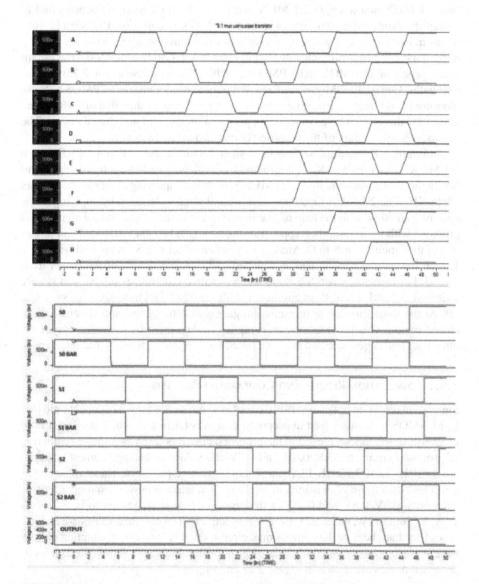

FIGURE 2.25   Input and output waveforms of existing 8:1 MUX design.

2.5 PROPOSED 20T DESIGN FOR 8:1 MUX

As concluded, a multiplexer designed by virtue of pass transistor logic (PTL) approach requires a minimal number of transistors resulting in an area efficient model. But by wielding this design style, power consumed by the circuit is higher and besides it does not reflect optimal performance in the output. However, the transmission gate logic (TGL) approach persuades lowering of power consumption of a circuit and furthermore it yields preferable performance at the output. Thereupon, proposed model of 8:1 MUX takes advantage of both the design styles by making use of merits of PTL in addition to TGL technique. As stated, 8:1 MUX can be implemented by utilizing two 4:1 MUX and a single 2:1 MUX and hence this procedure is incorporated to design the proposed model. The proposed 8:1 MUX is constituted of terribly minimum transistors as compared to conventional transmission gate design i.e. 16 pass transistors and 2 transmission gates, which makes total of 20 transistors. The circuit is comprised of 18 NMOS and 2 PMOS or rather 16 pass transistors and 2 transmission gates. Generally, NMOS is utilized as a pass transistor instead of PMOS because mobility of its charge carriers, i.e., electrons, is much higher than that of PMOS, i.e., holes, and therefore it helps in reducing propagation delay of any digital circuit. The transistor level diagram of the proposed circuit is delineated in Figure 2.26.

As inferred, 8:1 multiplexer can be designed with the aid of two 4:1 MUX and a 2:1 MUX. Upper 4:1 MUX is constructed by employing eight pass transistors N1 to N8. In the similar fashion, lower 4:1 MUX is built by employing eight pass transistors N9 to N16. The 2:1 MUX incorporated in second stage is designed by applying transmission gate logic with the help of two transmission gates. When control signal S2 is provided with logic zero, then upper transmission gate becomes activated and passes one of the inputs from A to D. Analogously, when select line S2 is applied with logic 1, then lower transmission gate is actuated and thus passes one of the inputs from E to H. For instance, when S2=S0=0 and S1=1 then pass transistors N5 and N6 are switched ON and acts as short circuit while all the other pass transistors are switched off. At the same time upper transmission gate is also triggered on and behaves like closed switch. This provides output as replica of data input C. Likewise, one of the other data inputs gets selected at a time based on the logic supplied to the select lines.

2.5.1 SIMULATION RESULTS AND COMPARATIVE ANALYSIS

Simulation is carried out using HSPICE tool at 45nm technology. The aspect ratio (W/L) of NMOS transistors used in proposed design is taken as 1 while the aspect ratio of PMOS transistors employed in the proposed circuit is taken as 3. After simulation of proposed circuit at 0.9V, 0.8V, and 0.7V, the values of leakage current acquired are 3.6201E-12A, 3.2201E-12A, and 2.8202E-12A respectively. These values are 8 times less than those of existing circuit. Similarly, leakage power acquired after calculation at 0.9V, 0.8V, and 0.7V for the proposed design comes out to be 3.2580E-12W, 2.5760E-12W, and 1.9741E-12W respectively. The obtained values are about 7 times less than those of the conventional circuit. Overall, power consumption of both the circuits are nearly equal in order of Nano watts. Input and output waveforms of proposed design are shown in Figure 2.27 and Figure 2.28 respectively.

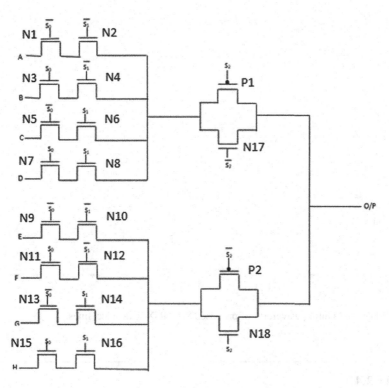

FIGURE 2.26   Proposed circuit of 8:1 MUX.

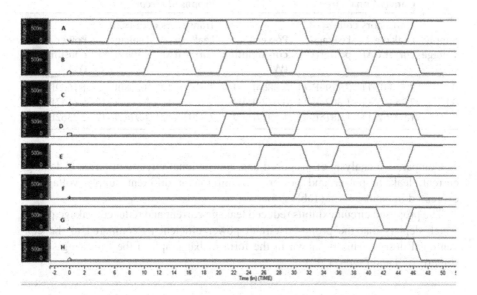

FIGURE 2.27   Input waveforms of proposed 8:1 MUX.

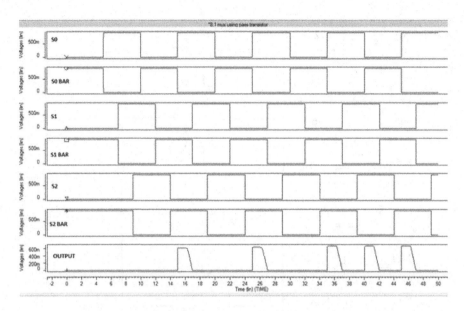

FIGURE 2.28 Output waveform of proposed 8:1 MUX with select lines.

TABLE 2.4
Comparative analysis of existing and proposed 8:1 Multiplexer

	Conventional circuit			Proposed circuit		
	Transistors count=28			Transistors count=20		
Supply voltage	Leakage current(A)	Leakage power(W)	Power consumption (W)	Leakage current(A)	Leakage power(W)	Power consumption (W)
0.9	2.5333E-11	2.2799E-11	7.8864E-09	3.6201E-12	3.2580E-12	1.1609E-08
0.8	2.2532E-11	1.8025E-11	3.6279E-09	3.2201E-12	2.5760E-12	5.3147E-09
0.7	1.9736E-11	1.3815E-11	1.5387E-09	2.8202E-12	1.9741E-12	2.2362E-09

Comparative analysis of proposed and conventional circuits in terms of leakage current, leakage power and power consumption at different supply voltages is expressed in tabular form (Table 2.4).

The proposed circuit exhibits reduced leakage current and reduced leakage power, which depicts that the proposed circuit power efficient as compared to the conventional design. This is shown in the form of bar graph in the following section (Figure 2.29 and Figure 2.30).

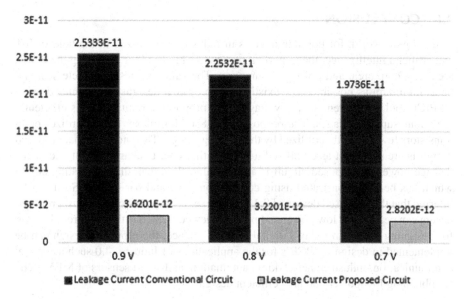

FIGURE 2.29   Comparison in leakage current of proposed and existing design.

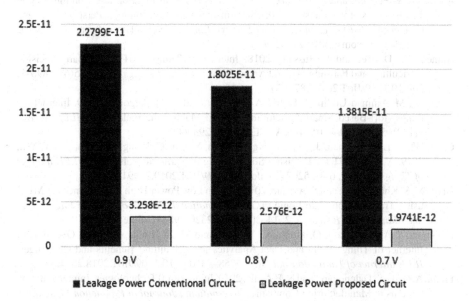

FIGURE 2.30   Comparison in leakage power of proposed and existing design.

2.6 CONCLUSION

In the digital world, for portable devices in Industry 4.0 utilized in the field of IoT and artificial intelligence, the important parameters required for designing the PCB's are delay, chip area, and power consumption. The value of these parameters should be as minimum as possible. The existing and proposed designs are simulated using HSPICE tool at different supply voltages. The proposed circuit performs efficiently at various supply voltages. The proposed 20T 8:1 Multiplexer design utilizes eight transistors fewer than that utilized by the existing design. This indicates that proposed design is area efficient as compared to the existing one. Leakage current and hence leakage power of proposed circuit is about 7 times less than that of existing circuit which has been demonstrated using comparison table and bar graphs. So it can be inferred that the proposed design of 8:1 Multiplexer is power efficient. Thus, proposed design can be used for low power and portable devices which is the basic requirement for designing PCB's in today's digital industry. Hence, the proposed design can be implemented in designing PCB's for the applications of Industry 4.0 such as optical communication, industrial networking, automation, and smart sensors of MEMS etc. for obtaining low power and area efficient designs.

REFERENCES

Aceto, G., V. Persico, and A. Pescape. 2019. A Survey on Information and Communication Technologies for Industry 4.0: State-of-the-Art, Taxonomies, Perspectives, and Challenges. *IEEE Communications Surveys Tutorials*, vol. 21, no. 4, pp. 3467–3501. doi: 10.1109/comst.2019.2938259.

Ahmed, N., D. De, and I. Hussain. 2018. Internet of Things (IoT) for Smart Precision Agriculture and Farming in Rural Areas. *IEEE Internet of Things Journal*, 5(6): 4890–9. doi: 10.1109/JIoT.2018.2879579.

Ayaz, M., M. Ammad-Uddin, Z. Sharif, A. Mansour, and E. M. Aggoune. 2019. Internet-of-things (iot)-Based Smart Agriculture: Toward Making the Fields Talk. *IEEE Access*, 7(129): 551–83. doi: 10.1109/ACCESS.2019.2932609.

Chen, W., Y. Lin, R. Chen, J. Liao, F. Ng, Y. Chan, Y. Liu, C. Wang, C. Chiu, and T. Yen. 2019. Agritalk: IoT for Precision Soil Farming of Turmeric Cultivation. *IEEE Internet of Things Journal*, 6(3): 5209–23. doi: 10.1109/JIoT.2019.2899128.

Dixit, A., S. Khandelwal, and S. Akashe. 2014. Design Low Power High Performance 8:1 MUX using Transmission Gate Logic (TGL). *International Journal of Modern Engineering and Management Research*, 2: 14–20. ISSN: 2320-9984 (Online).

Elijah, O., T. A. Rahman, I. Orikumhi, C. Y. Leow, and M. N. Hindia. 2018. An Overview of Internet of Things (IoT) and Data Analytics in Agriculture: Benefits and Challenges. *IEEE Internet of Things Journal*, 5(5): 3758–73. doi: 10.1109/JIoT.2018.2844296.

Frank, A. G., L. S. Dalenogare, and N. F. Ayala. 2019. Industry 4.0 Technologies: Implementation Patterns in Manufacturing Companies. *International Journal of Production Economics*, 210: 15–26. doi: 10.1016/j.ijpe.2019.01.004.

Giorgetti, A., Lucchi, M., Tavelli, E., Barla, M., Gigli, G., Casagli, N., and Dardari, D. 2016. A Robust Wireless Sensor Network for Landslide Risk Analysis: System Design, Deployment, and Field-Testing. *IEEE Sens. J.*, 16: 6374–86. doi: 10.1109/JSEN.2016.2579263.

Gupta, I., N. Arora, and B. P. Singh. 2012. New Design of High Performance 2:1 Multiplexer. *International Journal of Engineering Research and Applications (IJERA)*, 2(2): 1492–6. ISSN: 2248-9622

Javaid, M., A. Haleem, R. Vaishya, S. Bahl, R. Suman, and A.Vaish. 2020. Industry 4.0 Technologies and Their Applications in Fighting Covid19 Pandemic. *Diabetes and Metabolic Syndrome: Clinical Research and Reviews*, 14(4): 419–22. doi: 10.1016/j.dsx.2020.04.032.

Kovacs, I. and I. Husti. 2018. The Role of Digitalization in the Agricultural 4.0 – How to Connect the Industry 4.0 to Agriculture? *Hungarian Agricultural Engineering*, 38–42. doi: 10.17676/HAE.2018.32.38.

Liu, Y., X. Ma, L. Shu, G. P. Hancke, and A. M. Abu-Mahfouz. 2020. From Industry 4.0 to Agriculture 4.0: Current Status, Enabling Technologies, and Research Challenges. *IEEE transaction on Industrial Informatics*, 1–13.doi: 10.1109/TII.2020.3003910.

Metzgen, P., and D. Nancekievill. 2005. Multiplexer Restructuring for FPGA Implementation Cost Reduction. *Anaheim, California, USA DAC*, 421–6. doi: 10.1145/1065579.1065692.

Miranda, J., P. Ponce, A. Molina, and P. Wright. 2019. Sensing, Smart and Sustainable Technologies for Agri-food 4.0. *Computers in Industry*, vol. 108, pp. 21–36. doi:10.1016/j.compind.2019.02.002.

Mishra, M. and S. Akashe. 2013. High Performance, Low Power 200 Gb/s 4:1 MUX with TGL in 45 nm Technology. *Journal of Applied Nanoscience, Springer*, 4(3): 271–7. doi: 10.1007/s13204-013-0206-0.

Oztemel, E., and S. Gursev. 2020. Literature Review of Industry 4.0 and Related Technologies. *Journal of Intelligent Manufacturing*, 31(1): 127–82. doi: 10.1007/s10845-018-1433-8.

Qi, Q., and F. Tao. 2018. Digital Twin and Big Data towards Smart Manufacturing and Industry 4.0: 360 Degree Comparison. *IEEE Access*, 6, 3585–93.doi: 10.1109/ACCESS.2018.2793265.

Raut, R. D., A. Gotmare, B. E. Narkhede, U. H. Govindarajan, and S. U. Bokade. 2020. Enabling Technologies for Industry 4.0 Manufacturing and Supply Chain: Concepts, Current Status, and Adoption Challenges. *IEEE Engineering Management Review*, 1–1. doi: 10.1109/EMR.2020.2987884.

Rojko, A. 2017. Industry 4.0 Concept: Background and Overview. *International Journal of Interactive Mobile Technologies*, 11: 77–90. doi:10.3991/ijjm.v11i5.7072.

Roohi, A., H. Khademolhosseinei, S. Sayedsalehi, and K. Navi. 2011. A Novel Architecture for Quantum-Dot Cellular Automata Multiplexer. *International Journal of Computer Science*, 8(6): 55–60. ISSN (Online): 1694-0814.

Rubmann, M., M. Lorenz, P. Gerber, M. Waldner, J. Justus, and P. Engel. 2015. Industry 4.0: The Future of Productivity and Growth in Manufacturing Industries. *BCG Online Article*.

Rüttimann, G. B. 2016. Lean and Industry 4.0—Twins Partners or Contenders? *Journal of Service Science and Management*, 9. doi: 10.4236/jssm.2016.96051.

Sarita, J. H. 2013. Design and Implementation of Low Power 4:1 Multiplexer using Adiabatic Logic. *International Journal of Innovative Technology and Exploring Engineering (IJITEE)*, 2(6): 224–8. ISSN: 2278-3075.

Suzuki, T., Y. Kawano, Y. Nakasha, S. Yamaura, T. Takahashi, K. Makiyama, and T. Hirose. 2007. A 50-Gbit/s 450-mW Full-Rate 4:1 Multiplexer with Multiphase Clock Architecture in 0.13-µm InP HEMT Technology. *IEEE Journal of Solid State Circuits*, 42(3): 637–46. doi: 10.1109/JSSC.2006.891495.

Xu, L. D., E. L. Xu, and L. Li. 2018. Industry 4.0: State of the Art and Future Trends. *International Journal of Production Research*, 56(8): 2941–62. doi:10.1080/00207543.2018.1444806.

Zambon, I., M. Cecchini, G. Egidi, M. G. Saporito, and A. Colantoni. 2019. Revolution 4.0: Industry vs. Agriculture in a Future Development for SMEs. *Processes*, 7(1): 36. doi: 10.3390/pr7010036.

3 Data Science and AI for E-Governance

A Step towards Society 5.0

Amandeep Kaur, Neerja Mittal, and P. K. Khosla

CONTENTS

3.1 INTRODUCTION

In the world of Industry 4.0, data science and artificial intelligence have emerged to be strong forces in shaping the solutions to various problems encountered by governments and societies. The current pandemic situation due to Covid-19 has also witnessed the use of such technologies to track the patients, trace the source of transmission, and enforce social distancing and lockdown norms etc. Data about the disease is being collected from the hospitals and shared with various researchers all over the globe in order to develop countermeasures to safeguard from the disease. These datasets are also available on online learning platforms such as Kaggle.

The most famous example of the penetration of artificial intelligence into today's world is that of Chatty Gargoyle (Logistics Insider 2019) installed at Denver International Airport, which is a perfect blend of creativity and technology. Such initiatives are being taken by various airports around the world to enhance the user experience by using artificial intelligence to build chatbots, robotics for transportation services, biometrics for facial recognition and virtual reality and augmented reality for aviation related trainings (Atkinson, Baroni, Giacomin, Hunter, Prakken, Reed, and Villata 2017). There are numerous data science challenges posted on various online platforms based on real-life problems such as climate change, wildlife, human psychology, poverty, food security, etc. Open data is available for these challenges and optimum solutions are submitted by researchers. Data science plays important role during election of new governments, analysis of government policies, understanding delays in processes and various other activities related to government functioning (Jimenez, Solanas, and Falcone 2014). Therefore, data science holds a lot of potential to serve the society for the better good. However, artificial intelligence and data science have not entered the mainstream of populations in most developing countries. Thus, the exploitation of the huge potential of these technologies depends upon the widespread integration of real-life processes in various domains with these emerging areas.

The relationship between artificial intelligence and data science is shown in Figure 3.1. Data science deals with everything that can be done with data such as data collection, data pre-processing, data cleaning, data processing, data analysis, and data visualization, whereas artificial intelligence deals with making computers mimic the behavior of humans in form of responses for certain situations. The people who deal with the domain of data science are called data scientists, who are responsible for working on different sizes, scales, and variety of data. They are also responsible for ensuring the safety and security of the information which is perceived by analysis of data. The privacy of data is also a major concern which they need to address. They need to be well versed with diverse domains such as statistics, mathematics, programming, and databases, as well as storytelling.

Artificial intelligence is a much broader concept which further comprises machine learning which is actually used by developers and analysts. Machine learning is further of different types: (i) supervised, (ii) unsupervised, and (iii) reinforcement. In supervised machine learning, input as well as output labels are provided to build and run models. In unsupervised machine learning, only input labels are provided and output is generated by machine. And in reinforcement machine learning, human input yields machine output and then again the reward/punish cycle continues to refine the model further. Machine learning further comprises a variety of techniques such as regression, classification, clustering, feature reduction, etc. And there are a number of algorithms under each of these techniques which operate upon different kinds of data such as linear regression, which applies to numerical data, and logistics which,

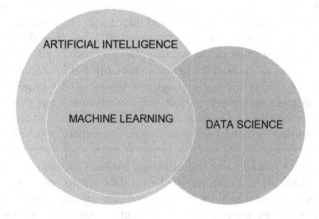

FIGURE 3.1   Relationship between artificial intelligence and data science (Pranav, 2018).

TABLE 3.1
Various aspects of machine learning

Category	Sub-category
Types of machine learning	• Supervised
	• Unsupervised
	• Reinforcement
Regression techniques	• Linear
	• Logistic
Classification techniques	• Neural Networks
	• K-Neural Networks
	• Decision Tree
	• Random Forest
	• Support Vector Machines
	• Naïve Bayes
Clustering techniques	• K-Means
	• Anomaly Detection
Feature reduction techniques	• T-Distrib Stochastic Neib Embedding
	• Principle Component Analysis
	• Canonical Correlation Analysis
	• Linear Discriminant Analysis
Evaluation/analysis parameters	• Bias Variance Tradeoff
	• Underfitting
	• Overfitting
	• Inertia
	• Accuracy
	• Precision
	• Specificity
	• Sensitivity

applies to categorical data. The information related to various aspects of machine learning has been provided in Table 3.1.

This chapter further discusses various data science and artificial intelligence applications which facilitate E-government and benefit the society.

3.2 DATA SCIENCE FOR E-GOVERNMENT AND SOCIETY

As the world is hit with Covid-19 pandemic, the activities which could be done by the physical presence of an individual observed a sudden halt due to imposed lockdowns globally. Be it governance or society, the digital space has emerged as a savior. Schools, colleges, and universities have started education services online. Hospitals have started providing e-prescriptions for normal routine checkups, thereby saving time and space for emergency services. The governments and related offices have started the discussions on various issues through video conferencing. The companies and industries have shifted their businesses online. Banking and healthcare are also taking advantage of digital technology. Therefore, with the use of digital space, data

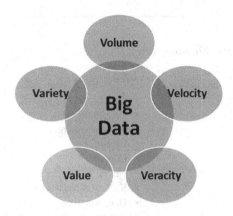

FIGURE 3.2   5Vs of big data.

is being generated in huge amount which can be explored and analyzed with the help of data science. With the rise of data, the term "big data" comes into picture which is characterized in terms of 5Vs: Velocity, Variety, Veracity, Volume, and Value, as shown in Figure 3.2. Governments are having heaps of information about their citizens which they could use to find out various trends and prospects. This information pool can be utilized by applying data analytics for a smarter future.

E-governance is basically defined as the use of Information and Communication Technology (ICT) by government to facilitate the collection and disbursal of information and services to citizens, businesses, society, and other government as well as non-government agencies in an efficient, cost-effective manner (Cath, Wachter, Mittelstadt, Taddeo, and Floridi 2018). There are usually four models of E-governance based on the services they provide: (i) Government to Citizens (G2C) (ii) Government to Employees (G2E) (iii) Government to Government (G2G) (iv) Government to Business (G2B) (Agbozo and Spassov 2018). The use of these online services tends to generate a huge amount of data which further takes the shape of big data as it comes from various sources and in different shapes, sizes, and formats. To analyze this data, various big data analytics tools have to be used. The traditional database management processing systems are unable to handle big data; therefore a number of big data technologies, tools, and techniques have emerged such as MapReduce programming model, Google File System, Big Table, Chubby Lock Service, Cassandra, Hadoop, Dremel, Pregel, Spanner, Shark, Spark, MegaStore, MLBase, NoSQL, HBase, Hadoop Distributed File System, Yet Another Resource Negotiator (YARN), Mahout, Chukwa, etc. The use of various data science and big data tools has led to transformation in the government sector by providing real-time solutions to various problems in agriculture, healthcare, transportation, and education, etc. (Navdeep, Arora, and Sharma 2016). The use of such tools and technologies also brings along a number of challenges related to privacy and security of the data involved. The cloud storage of data is also a concern as well as accuracy of the information perceived from the data. Data science plays a very important role in decision-making processes of organizations, because it provides valuable insights from available data.

For social benefit also, data science is playing a significant role. For instance, there are a number of social media platforms existing in today's world and each one has their own reason to be popular among the masses. The data about the user's choices, perspectives, decisions, likes and unlikes, faith, worship, and social circle is being generated and captured on daily basis. This data may provide insights about different user groups in different regions of the world. This data is said to be secure but still there is a possibility of unauthorized access which can never be ruled out, no matter how secure the platform is (Souma 2017). However, this has made the life of the people easier by providing them personalized information at the click of a button. Data-science-based recommendation systems are widely used by various shopping apps to provide better user experience and sell products with better efficiency.

3.3   ARTIFICIAL INTELLIGENCE FOR E-GOVERNMENT AND SOCIETY

Artificial intelligence mainly deals with the automation of processes related to E-governance and society. Machine learning is the widely deployed technique in the AI domain which can help in making predictions about the success of certain E-governance initiatives or other services for the benefits of the society (Makridakis 2017). Table 3.2 lists various areas of E-government and societal services along with the application of artificial intelligence in these. These AI applications are assisting in making the services reachable to the citizens as well as making them faster, efficient, reliable, and accurate.

Personalized recommendation techniques are one of the interesting applications of artificial intelligence which are being deployed to improve the quality of government-to-business online services. Computational modeling of argument is another emerging aspect of artificial intelligence research which is being used to analyze arguments and counterarguments during some debate or discussion. The interoperability of the platforms and tools used for E-government is an important area of research as citizens operate upon different systems and from different regions.

Fuzzy methods are being used in E-government as an approach to address security concerns (Elssied., Ibrahim, and Yousif 2011). A distributed and secure model is being used by China to provide information to the citizens over different platforms (Lu 2007). Therefore, the governance has become data driven in the present world and the widespread use of artificial intelligence has made it possible to analyze this accumulated data effectively and produce patterns which could assist in decision making.

3.4   TOOLS AND TECHNIQUES

Various data science and artificial intelligence tools which can be used to address real time issues in the area of E-governance and society are summarized in Table 3.3.

TABLE 3.2

Various E-government and societal services and AI applications (Guo and Lu 2007; Kaya 2019)

Sr. No.	Services	AI applications
1	Law enforcement	• Facial recognition • Real time Image processing and enhancement • Robotics (Smart bots) • Speech recognition • Credit card fraud analysis systems • ATM surveillance
2	Automation of routine tasks	• Natural language processing • AI assist robots • Credit card • Human resource management
3	Disaster management	• Food and medicine delivery services • Threat prediction and avoidance • Identifying vulnerable populations • UAV assisted information disbursal systems
4	Maintenance of public infrastructure	• Automated solar panel systems • Automated cooling systems • Internet of things enabled services • Cloud storage of data
5	Education	• Understanding student behaviours • Tools for skill development • AI solutions for course module development • Personalization and individual learning • Universal accessibility • Online tutoring services
6	Healthcare (Samantha M., 2020).	• Deep learning for diagnosing the diseases • Patient-doctor online interfaces • Chatbot services for psychological issues • Machine learning for disease prediction • Patient experience improvement using AI • Dementia detection
7	Defence	• Expert systems • Robotic equipments for precision and accuracy • Automated training services • Image enhancement • AI based defence startups • Smart logistics
8	Businesses and industries	• automated enterprise resource planning systems • 3D modelling and design • Human resource services • Smart solutions to address consumer complaints

TABLE 3.2 (Continued)
Various E-government and societal services and AI applications (Guo and Lu 2007; Kaya 2019)

Sr. No.	Services	AI applications
9	Transportation	• Self-driving vehicles • Environmental pollution reduction • Energy management • Planning and automation • Traffic management • Sensor based railway cargo services • Pattern recognition for road safety • Safe and clean transport

TABLE 3.3
Various data science and artificial intelligence tools for E-governance and society (Sayantini, 2019)

Category	Tools	Description
Data science (Data Flair, 2019)	SAS	• Proprietary software • Used for statistical modeling and analysis by large organizations • Supports various types of data formats • Employs data encryption algorithms
	Apache Spark	• Analytics engine for data science • Handles batch processing and stream processing • Provides in-memory computing and fault tolerance • Offers APIs programmable in Python, Java, R and Scala
	BigML	• Cloud computing based data science tool • Predictive modeling tool • Easy to use web interface using Rest APIs.
	D3.js	• Javascript library for interactive visualizations • Open source • Can be combined with CSS
	MATLAB	• Numerical computing environment • Simulates neural networks and fuzzy logic • Image and signal processing
	ggplot2	• Data visualization package for R • Can create maps such as choropleths, cartograms, hexbins

(continued)

TABLE 3.3 (Continued)
Various data science and artificial intelligence tools for E-governance and society (Sayantini, 2019)

Category	Tools	Description
	Tableau	• Data visualization software having powerful graphics • Enterprise software • Free version is called Tableau Public
	Jupyter	• Web application Data science tool that supports Julia, Python, R • Provides data cleaning, statistical computation, visualization • Does predictive modeling
	Matplotlib	• Plotting and visualization • Pyplot-open source alternative
	NLTK	• Natural Language Toolkit for natural language processing • Applications: parts of speech tagging, word segmentation, machine translation, text to speech speech recognition
	Scikit-learn	• Python library for machine learning algorithms • Supports data preprocessing, classification, regression, clustering, dimensionality reduction
	TensorFlow	• Used for advanced machine learning such as deep learning • Named after tensors (multidimensional arrays) • Runs on both CPUs and GPUs
	Weka	• Waikato environment for knowledge analysis tool for data mining • Open source GUI software
Artificial Intelligence (Guido, Titus, and Hichamel, 2017)	Heuristics	• Defined by Webster (heuristics definition) as involving or serving as an aid to learning, discovery, or problem-solving by experimental and especially trial and error methods • Ant colony optimizations and genetic algorithms fall under this category
	Support vector machines	• Provides classification such as whether the email is spam or not • Idea is to find the boundary line that separates into two classes
	Artificial Neural Networks	• For speech recognition, recurrent networks are used

TABLE 3.3 (Continued)
Various data science and artificial intelligence tools for E-governance and society (Sayantini, 2019)

Category	Tools	Description
x		• Machine learning technique loosely modeled after the neural structure of a brain
	Markov decision process	• Framework for decision-making modelling
		• Example of application: inventory planning
	Theano	• Abnormal state neural systems library
		• Can be used to run complex models due to high speed
	Caffe	• BSD-authorized C++ library with Python Interface
		• Google's Deepdream depends on this framework
	MxNet	• Opensource scalable framework
		• Supports multi-GPU training
	Keras	• High level Python library for neural networks
		• Used for image recognition
	PyTorch	• AI system created by Facebook
		• Used for image recognition
	CNTK	• combines various model types such as feed-forward DNNs, convolutional nets (CNNs), and recurrent networks (RNNs/LSTMs)
		• Implements stochastic gradient descent learning
	Auto ML	• Used for optimizing machine learning tasks
		• Contains a number of tools for machine learning
	OpenNN	• Provides advanced analytics
		• It provides a tool called Neural Designer for advanced analytics which provides graphs and tables
	H20: Open Source AI platform	• Deep learning platform
		• Applications: predictive modelling, risk and fraud analysis, insurance analytics, advertising technology, healthcare and customer intelligence
	Google ML Kit	• Machine learning kit for Android/IOS mobile developers
		• Applications: face and text recognition, barcode scanning, image labelling

3.5 FUTURE PROSPECTS AND TRENDS

As the present world is engulfed in the Covid-19 pandemic, the digital technologies are turning out to prove beneficial for this physically disconnected world.
The people can sit in their comfort zones and still achieve a lot by exploiting the
domains of data science and artificial intelligence. These technologies are being
widely deployed in almost all spheres of society and E-governance to assist in the
day to day operations. There are some future prospects and challenges in these
areas which need to be pondered upon such as the following, leading to several
research directions.

- The revolution Industry4.0 deals with the interconnection of people, processes
  and devices using the idea of Internet of Things. This concept is still in its
  infancy as it has not penetrated much into the physical world due to certain
  limitations. Further research can be done in this area to enable people to control
  all the devices in their home with a single click as well as provide connectivity
  to their office environment even in their physical absence.
- The storage, management, and analysis of big data are other challenges
  which need thorough research. The existing tools are not capable enough to
  capture the value of big data and database management systems also need
  to be enhanced in capability in order to store and bring out a meaning of
  this data.
- In E-government services, the data collected is about citizens. Therefore, the
  data needs to be captured and stored using secure and privacy based models.
- In e-healthcare services using artificial intelligence, security, and privacy of
  patient data is a major concern.
- The ethical, legal, and societal implications of artificial intelligence should be
  addressed with extreme measures.
- There should be methods to analyze dark data, which is being accumulated but
  not used for any purpose.

3.6 CONCLUSIONS

It is concluded that various data science and artificial intelligence applications hold
a lot of potential to shape the E-government and societal services into fast paced and
more efficient processes. However, there are numerous challenges that come along
which need to be addressed well before allowing these technologies to penetrate
deeply into the systems. The misuse of data generated using these technologies could
lead to conflicts around the world as the weapons and missile systems are also making
use of advanced technologies. There are a number of tools available in the areas of
data science and artificial intelligence which have been listed in this chapter. These
tools can be used for a variety of applications depending upon the kind of problem
which we tend to solve. The balanced and careful use of these technologies could
really help in transforming this physically disconnected world due to Covid-19 pandemic into a digitally connected world.

REFERENCES

Agbozo, E., and Spassov, K. 2018, April. Establishing Efficient Governance through Data-Driven E-government. In *Proceedings of the 11th International Conference on Theory and Practice of Electronic Governance* (pp. 662–4).

Atkinson, K., Baroni, P., Giacomin, M., Hunter, A., Prakken, H., Reed, C., and Villata, S. 2017. Towards Artificial Argumentation. *AI Magazine*, 38(3): 25–36.

Cath, C., Wachter, S., Mittelstadt, B., Taddeo, M., and Floridi, L. 2018. Artificial Intelligence and the 'Good Society': The US, EU, and UK Approach. *Science and Engineering Ethics*, 24(2): 505–28.

Data Flair. 2019. *14 Most Used Data Science Tools for 2019 – Essential Data Science Ingredients*. Retrieved from https://data-flair.training/blogs/data-science-tools/.

Elssied, N. O. F., Ibrahim, O., and Yousif, A. 2011. Security in E-government Using Fuzzy Methods. *International Journal of Advanced Science and Technology*, 37: 99–112.

Guido, D., Titus, S. E., and Hichamel, B. 2017. *Part 2: Artificial Intelligence Techniques Explained: Zooming in on Fundamental AI Techniques*. Retrieved from www2. deloitte.com/nl/nl/pages/data-analytics/articles/part-2-artificial-intelligence-techniques-explained.html.

Guo, X., and Lu, J. 2007. Intelligent E-government Services with Personalized Recommendation Techniques. *International Journal of Intelligent Systems*, 22(5): 401–17.

Heuristics Definition. Retrieved from www.merriam-webster.com/dictionary/heuristic.

Jimenez, C. E., Solanas, A., and Falcone, F. 2014. E-government Interoperability: Linking Open and Smart Government. *Computer*, 47(10): 22–4.

Kaya, T. 2019, October. Artificial Intelligence Driven E-government: The Engage Model to Improve e-Decision Making. In *ECDG 2019 19th European Conference on Digital Government* (p. 43). Academic Conferences and Publishing Limited.

Logistics Insider. 2019. *Chatty Gargoyle Greets Passengers at Denver International Airport*. Retrieved from www.logisticsinsider.in/chatty-gargoyle-greets-passengers-at-denver-international-airport/?nonamp=1.

Lu, X. 2007, July. Distributed Secure Information Sharing Model for E-Government in China. In *Eighth ACIS International Conference on Software Engineering, Artificial Intelligence, Networking, and Parallel/Distributed Computing (SNPD 2007)* (Vol. 3, pp. 958–62). IEEE.

Makridakis, S. 2017. The Forthcoming Artificial Intelligence (AI) Revolution: Its Impact on Society and Firms. *Futures*, 90, 46–60.

Navdeep, P., Arora, M., and Sharma, N. 2016. Role of Big Data Analytics in Analyzing e-Governance Projects. *New Trends in Business and Management: An International Perspective*, 6(2) (April_June), 53–63.

Pranav, G. 2018. *The Subtle Differences among Data Science. Machine Learning, and Artificial Intelligence*. Retrieved from https://medium.com/@prnvg/the-subtle-differences-among-data-science-machine-learning-and-artificial-intelligence-45d4208c61f.

Samantha, M. 2020. *Challenges of Artificial Intelligence Adoption in Healthcare*. Retrieved from https://hitinfrastructure.com/news/challenges-of-artificial-intelligence-adoption-in-healthcare.

Sayantini. 2019. *Top 12 Artificial Intelligence Tools and Frameworks You Need to Know*. Retrieved from www.edureka.co/blog/top-12-artificial-intelligence-tools/.

Souma, D. 2017. *Data, Digitisation, and Digital India: How Data Analytics Can Facilitate Smoother E-governance*. Retrieved from https://analyticsindiamag.com/data-digitisation-digital-india-data-analytics-can-facilitate-smoother-e-governance/

REFERENCES

4 Application Areas of Data Science and AI for Improved Society 5.0 Era

Sneha Mishra, Priya Porwal, and Dileep Kumar Yadav

CONTENTS

4.1 INTRODUCTION

Artificial intelligence and data science being the effective and essential technologies in this modern era of technology play a much vital role to the society. The

E-government system needs to be improved for the convenience of the society; simultaneously a great work is to be done with data handling that too needs better technology. The management of data, the extraction of useful information through the data, the better analysis of the data, and digging out the best from the data and performing other activities more intelligently nowadays. The activities must save time be secure and easy to handle. Data science and artificial intelligence are the essential pillars to improve human life. The necessity of the hour is to perform the robust activities with an ease and to reduce the complexity of the work. Data science (Schwartz 2016) deals with data which is limitless and AI on the other hand solves complex problems as intelligently as human brain, in some cases even better than the human brain; for instance in retail industry the chances of human error are very obvious in calculations but a machine can do much better.

Society needs to deal with n number of data, and that n number of data is also increasing at such a fast pace that dealing with that data statistically is becoming more complex day by day. If we take an example of social media, an image which is shared by a person in a specific group or broadcast which contains a hundred or even more than hundred people, within a microsecond that hundred-people group has linked with another hundred-people group, forwarded that message to other people and those people have forwarded to further people, and so on. The multiplicity of the data within seconds increases exponential times. This is a very small example of the massiveness of increase in the data. In many areas like medical sciences the data is increasing at such a speed that to maintain the authenticity and extraction of best knowledge from the data has become the great reason for the major employment in various sectors of society these days. Various industries are training their employees to handle complex data and to draw signified patterns from the data. E-government is the combination of too many public sector organizations by using information and communication technologies and the interaction of these two bodies (public sector and the government). E-government deals with improvement of government processes, the connection of citizens with government through the Internet or online platform, and also to build external interaction which is called E-society. To deal with E-society various artificial intelligence technologies and machine learning techniques need to be employed by government organizations to improve E-government or interaction of government through society or public sector groups, so as to change or improve the way of communicating or engaging society and presenting the government's policies or works. This will somehow look like good governance. The transparency of the works of government will also be clear to organizations and bodies working under E-government. Data science and AI will collectively play a major role to build up the concept of E-government and society.

AI and data science will be needed to solve the robust situations of handling complex data related to both government and society; government and society are interrelated terms. Many areas in our society such as public safety, education, revenue, health and social sciences, labor, environment, transportation, program evaluation, strategic planning, etc. are the current applications of e- governance and these may have massive amount of data to be dealt with data analytics and artificial intelligence. Various machine learning techniques like neural network, natural language processing etc. will be used to implicate the task and for better handling of services. So in this chapter we will learn how our society and various public groups like healthcare, education, industries, agriculture, and social media may be a better version with the use of data science and artificial intelligence. *"Necessity is the mother of invention."* For the practice of E-governance the need came for the better management of data, data mining, data analytics, and machine learning and so artificial intelligence.

This chapter is organized as follows: In Section 4.2.1 you will learn how data science and society are interrelated with respect to data. The process of data science is covered in section 4.2.2 and applications of data science are described in Section 4.2.3; E-governance being the major part of society is explained in Section 4.3.1; the methods to be followed for better governance are outlined in Section 4.3.2 and further the application areas of E-governance are specified in Section 4.3.3. Section 4.4 describes the various platforms of new emerging technologies that ignite society in a better way. Sections 4.4.1, 4.4.2, 4.4.3 describe the Internet of things, artificial intelligence, and blockchain respectively. Section 4.5 covers the groups of society which use AI and data science with the uses and trends and applications in various sectors of the community like healthcare, education, communication, banking, etc. Section 4.6 summarizes the chapter.

4.2 DATA SCIENCE AND SOCIETY

In the past we used to memorize or write telephone numbers in our diaries, then the time came when mobile phones came into existence and people used to save contact numbers on their phones, but these days' smartphones store a lot of contact details, images, videos, applications, documents, etc. We carry our data in our hands, on desktops, refrigerators, microwaves, washing machines, dishwashers, and even electric bulbs and fans are controlled by smart devices and are creating a lot of data every day. The data is generated at an unstoppable pace in the cloud, and everywhere. Data is stored in electric appliances in our homes which we can control with our smartphones. Unimaginable data is coming daily, Terabytes and petabytes of data are generated by us, for which we need better processors and complex algorithms to process our data (Agnihotri, 2015). Earlier even our desktop was not that advanced to support our big data, but as time is passing better processing units are available. So we do this with data science. Data has a lot of power and we need to know how to process it. Data scientists are like detectives of data who spend a lot of time analyzing the data and enormous data sets. Analyzing data, applying machine learning techniques to data, and making sense out of that data, are executed by data science. The methods that let the data be visualized in their best and most understandable format is also done by using data analysis. Data analysts take accountability of presenting the data in the best possible way.

4.2.1 DATA SCIENCE AND SOCIETY INTERRELATION

Data science initially emerged with the principal of necessity for genuine applications, but later became the exploration area for the computer science department to investigate and analyze patterns of data, to improve business aspects by seeing the insights of the data and digging out certain hidden layers from the data. Business analytics use data science to give understanding to business and derive the figures of future execution of that firm. A data scientist studies the connections of consumers and sales by association rules and is thus able to increase the sale of any product. In data science the data is processed and analyzed in order to find out the hidden patterns and find out the insights of data. This helps in better decision-making abilities, business problems can be solved by this. Data science applies statistical theories for the extraction and processing of data; data wrangling is done to clean the data, exploring the data with different trends and outliers, then applying machine learning techniques like KNN (K-nearest neighbors), k-means, etc. are all the major roles played to form a better society. Davenport and Harris (2007) state that firms like Google and Facebook which are called digital firms have been the very early adopters in more traditional sectors like retail, finance, and transport. Some sectors like governments, hospitals, schools, transport (metro, train), and others have been very challenged by current developments, but are moving forward with mixed success (McKinsey 2011). Despite opportunities there are lots of challenges and hurdles with improved data governance and availability. For this the right skills and support of better leadership are required and those are the data scientists and data analysts.

4.2.2 PROCESS OF DATA SCIENCE

Usually when data is collected it is a raw set of data; unstructured data is a great mess that requires a lot of time and effort to convert into structured data. Data processing needs lot of time in data cleaning; there is a lot of unnecessary data, missing values, corrupted values, and inconsistent values. After cleaning data, we can move forward to model the data and then evaluate the data for which machine learning techniques will be required. Then comes the part of data analysis where a large amount of time is spent on applying predictions regarding the product.

Let us take an example of a company, any car which is newly launched in the market, using the previous data about the sales of their cars; they want to predict the category of people who might be interested in buying that car. For this the company needs to find the factors which made people more interested regarding the purchase of the product. The factors may be gender, age, estimated salary, etc., then

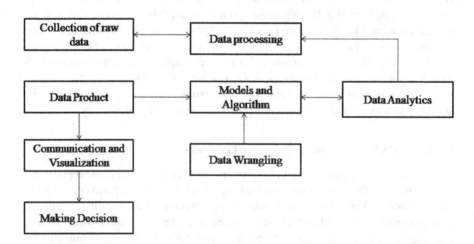

FIGURE 4.1 Steps of data processing.

applying regression to train the model so that we may define dependent and independent variables. Thus we predict the rate of purchase of the car, we can also calculate the accuracy by matrix method or just manually by confusion matrix. Thus in a similar manner there are many products and companies and even governments using data science. We can find dozens of such applications of machine learning and data science being used in real life, like getting recommendations of similar products while e-shopping. So after collection of data, data processing comes into existence (Shreshtha, Singh, Sahdev, Singha, and Rajput 2019).

Data science is about organizing data, assembling data and transferring data for further fusion. Below are the steps or channels associated with the processes of data science.

1. Data wrangling or data cleaning

Data wrangling is the most important step in the data process, the quality of the data is directly proportional to the accuracy of the data analysis. For good quality of data we need gathering of data, filtering the data, conversion of the data, exploring of data, and then integrating the data. The steps in data wrangling are as follows:

- Gather: Gathering of data is about extracting, parsing, and scraping the data.
- Filter: From raw data to something usable, remove corrupt data, remove invalid data combinations, and keep only what you may require.
- Conversion: Conversion of that data by applying preliminary filtering and scrubbing in some common formats like CSV, JSON, XML, SQL, etc.
- Exploring: Pre-analysis, preliminary queries, informs hypothesis.
- Integration: From schema-less to normalized, concatenation, joining, aggregating, and adding up the pieces, progressing towards big data.

2. Data investigation

Investigating the data is about evaluating, transforming, and presenting the same with an objective. It is about finding the obscure connection and exploring data. Data investigation leads to expulsion of the importance and beneficiary finishes from the massive volume of data by reading the diverse amount of data.

3. Data transfer

This is about transferring the data which fuses the system to change the numerical data from the facts into the edge that can be adequately grasped by the analyst who requires it. The exchange of data permits the change starting with one perspective to another.

4.2.3 APPLICATION AREAS OF DATA SCIENCE

Data science can be used to help the life of everyday people. Expectation, security, computer vision, and natural language processing are the core applications of the data science. Following are the application areas and their basic information with few examples in the real life scenarios. Table 4.1 describes the application areas of data science.

TABLE 4.1

Application areas of data science with their description.

Applications	Examples	Description
Internet search	Google, Yahoo, Bing	Make use of data science algorithms to deliver the best results for our queries less than the seconds.
Targeted advertisement	Netflix, Shein	Digital advertisement are decided by using data science algorithms, they can be targeted by users past behavior.
Website recommendation	Amazon, Twitter, Google play	Suggestions of similar products, adds lot of user experience, companies use this system or engine to promote their product with to users interest or relevance.
Image Recognition	Facebook, Whatsapp web	Upload the image and we get the recommendation of tagging the friends, this automatic tag suggestion feature uses face recognition algorithms.
Speech Recognition	Contana, Google voice, Siri	These applications work like we even don't have to type the message, just have speak out the message and it will convert into the text.
Airline Route Planning	Airline industries	Companies were struggling to maintain their occupancy, then airline companies started using data science to identify the strategic areas of airlines, they can predict delay, costumer experience, decision making and execution.
Fraud and risk detection	Industries,	The origin of data science came from finance discipline, companies use data science practices to rescue from losses, they started punching the products according to the costumers purchasing power.
Medical sciences	X-ray, CT-scan, Drugs production companies	Medical images analysis is the major area of data science algorithms and applying deep learning to provide better diagnostic activities. Virtual assistance of patients can bring basic health supports by describing our symptoms, it saves time, and generates the alarm whether it is serious to go to doctor or can be cured by this much effort.
Gaming	Pokémon go, Zingo, ES sports	Motion gaming the opponent analysis previous moves and decides the further steps this may not be possible without data science.

4.3 E-GOVERNMENT

Government industries can derive a significant benefit from the area of data science and analytics. Most of the government firms are carrying forward the business rules for obedience and suitability of the systems, even decision making and prediction analysis are more favored these days for government. Collaborative works of data science and AI have advanced the state-of- art results in several numbers of domains. Data science platforms lead to exploration of data into very deep to discover the better decisions for future. Data analytics focus explicitly on predicting future performances and obtaining the goals of performances and variations identification. The general idea of E-governance is to make processes transparent and open to maximum number of possible extent for the population or a particular beneficiary of that country, so is required for easier identification of the organization or cooperation. The transparency will lead to identification in the organization that who is responsible for the loss, corruption, and contribution in achieving the goals.

Further discussion will lead to roles of E-government in the society and application areas of E-governance. The major role of artificial intelligence technology in automating and facilitating E-government services for the development of society is sketched additionally.

4.3.1 ROLE OF E-GOVERNANCE IN SOCIETY

E-government is the application of information communication and technologies to run effective governance which shares information between common people, government, business etc. The E-governance is the diverse area and each and every age group in society is making use of e-governance aspects; for example many sectors like rail reservation and gas subsidy are related to E-government. Figure 4.2. denotes the applications of E-government being used by us in our everyday life.

4.3.2 E-GOVERNMENT SERVICES WITH ARTIFICIAL INTELLIGENCE

Artificial intelligence came into existence a few decades back in various complicated systems in theoretical form. But the advances in the computational powers and with a large size of data (big data) have lead AI to achieve outstanding results in vast number of domains. Computer vision (He 2016), medical sciences (Zhang 2018), natural language processing (Venugopalan 2015), and many other domains are being advanced by AI. Artificial intelligence is the ability of the machine to imitate the intelligence of human activities or behavior while improving its own performance; this describes the brain of the machine. AI performs many sophisticated jobs, such as driving a car and playing a game. E-government is the application of information and communication techniques to advance, exchange, and present government services for citizens and business to achieve the goals of better economy and advanced economy of citizens, industry, and government. The achievement of E-government is to better the quality and efficiency of government services. Implementing E-government services or applications can foster advantages like transparency, trust, citizen participation, environment support, etc.

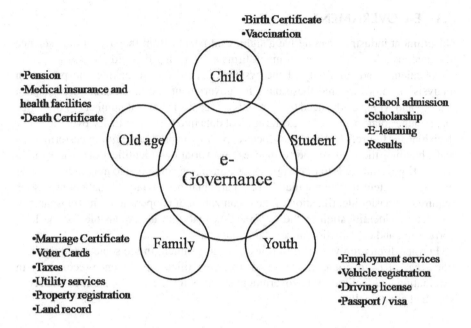

•Birth Certificate
•Vaccination

•Pension
•Medical insurance and
health facilities
•Death Certificate

•School admission
•Scholarship
•E-learning
•Results

•Marriage Certificate
•Voter Cards
•Taxes
•Utility services
•Property registration
•Land record

•Employment services
•Vehicle registration
•Driving license
•Passport / visa

FIGURE 4.2   Roles of E-governance at different stages of life in society.

TABLE 4.2
Comparison of EGDI and OSI in continents

Continent	E-government developing index	Online service index
Europe	0.77	0.79
America	0.59	0.61
Asia	0.57	0.62
Africa	0.34	0.36

But implementation of E-government services may face a lot of challenges like trust, lack of experts, inaccessibility, security, etc. These issues or challenges can be overcome with AI techniques and algorithms. If we look at the performance of E-government industry in several countries around the world, the European Union is leading in implementing E-government applications followed by the USA and Asia; Africa is located at the bottom due to its poor infrastructure. The UN-E-Government-Survey, 2018 reflects the lack of infrastructure at low level of E-government development index (EGDI) of countries. Table 4.2 displays the comparison of EGDI and online service index (OSI) according to region.

With this survey the investigation shows that human resources and technical infrastructure are the major and fine factors which play a key role in performing advanced development and providing good efficiency of E-government services.

A massive amount of data is generated by n number of sources which are homogenous in nature related to E-government and their services. So management of this data or information resources plays a crucial part in E-government services. This leads to a focus on utilization of various techniques AI, IoT, big data. AI technologies are one of the very active areas of this era to accept the challenges and give a better performance to the country to lead to higher development index in the world. The smart E-government platform assists and guides the efficient implementations of AI integrated E-government services. There are three layers which act as a link between E-government systems and proposed platform:

- **Smart contact citizens' service layer**: This acts an interface between the E-government services and citizens, and by intelligent agents, and then manages the current services.
- **Security layer:** This is responsible for implementation of strict policies and security measures to assure the privacy and security of data sharing applications and services.
- **Functional layer:** This layer is to maintain the accountability of the above two layers, it provides the core functionality and provides E-government automatic services. This layer is for analysis, planning, evaluation, and execution of the services.

4.3.3 APPLICATIONS OF E-GOVERNANCE

Often two terms, E-government and E-governance, which are the application of ICT for delivering government services, are used interchangeably. They provide improved public services, democratic processes, and strengthen support for the public. E-governance (Agbozo, 2019) offers a variety of services to citizens in an economical manner to better the relationship between the government and citizens using the latest technologies: payment of taxes and services by smart devices or online granting of identity cards, passports, issue of driving license, address proofs. Applying for granting and services are no longer face to face completion processes. E-government citizens have the ease of doing all such things remotely and instantly.

Across the globe, European countries, the United States, and Asia are ranked top in the development as these have good computing skills and better Internet access. Application areas of E-government on the basis of analysis of data are provided in Table 4.3.

4.4 EMERGING TECHNOLOGIES

Data science technology has grown rapidly over the past few years. AI and machine learning have now become the slanting professions for many individuals to follow. With such a significant number of organizations coming up and clamoring for the best new ability, today business is encountering a deficiency of skilled and qualified experts.

Through modern technology like neural network and machine learning, organizations universally are putting resources into instructing machines to "think"

TABLE 4.3
Application areas and role of data analytics.

Application areas	Role of data analytics
Revenue -taxation assessment -fraud analysis	• Assurance of accurate and equitable assessment. • Taking a record of those whom may have greater propensity of committing fraud.
Education -administration -student tracking	• Meeting mandatory programs and testing and maintaining the standards. • Collecting and discovering the factors working for success of students.
Transportation -accident reporting -road maintenance -route planning	• Examining and analyzing the factors heading accidents. • Prediction of road repair. • Planning the new effective routes for controlling the traffic flow.
Health and Medical Sciences -disease tracking -utilizing medical aid facilities -disease prevention	• Tracking the record of disease occurrence. • Creating profiles of frequent requires of medical aids and services. • Identification of risk population and head to needful intervention.
Public Safety -crime analysis -court sentencing analysis -Probation analysis	• Analyzing the type of crime and places of crime occurrence. • Tracking the record of all the cases processed in the court efficiently. • Analyzing and tracking the effectiveness of probation programs.
Environment -eco-system analysis -water /air quality testing	• Analyzing and understanding the factors helpful for healthy Eco-system. • Predicting and ensuring the maintained standards of water and air quality.

increasingly like people. Some of these technologies, such as IoT, AI, and blockchain, are as follows.

4.4.1   IoT (INTERNET OF THINGS)

IoT plays a significant role in society as well as E-governance. IoT can be defined as a term for a situation where anything might be embedded in a system, be remarkably distinguished, and collaborate with minimum human effort. There are several devices in the IoT world. Some devices have high level architecture with enormous memory limits and some have low level architecture with limited memory. Cloud computing is relied upon to assume a significant role in the IoT system. With cloud computing, IoT processing limits can be expanded in a versatile way. In addition, sensors can be utilized all over the place and handling sensor information can be practiced through cloud computing services.

Internet of Things is the connection of interrelated computing devices embedded with software, sensors, and electronics over a network so as to collaborating, transfer, and collect data. If we take the examples of our mobile phones, there are features like GPS tracking, face detection, adaptive brightness, voice detection, and many more, when they come together by interaction; they make a better system to anything which they tend to provide. In our homes all the things like AC, locks, lights, etc. can be managed on the same platform that is our mobile phone through the Internet. The Internet of Things is the major technology these days that can help all other technologies reach their complete potential. IoT can connect various things to a common platform, analyze the data piled, and then use that to build various analyses and can integrate different models to improve user experiences.

IoT devices are a set of devices that are connected to each other and exchange data through the Internet. IoT is the latest technology used for healthcare structure. IoT devices enables customers to decrease prosperity-related threats and human services costs by assembling the patient's subtleties and examinations data to sharing data stream using cloud computing. This structure involves a cloud-based technique for a motorized disease prescient system that uses sensors to measure various parameters of the patient like heart rate, temperature, etc. The purpose of such a system is in rapidly detecting disease from parameters such as patient's body temperature and blood pressure.

A. Features of IoT

Connect: Connect various things to IoT platform needs virtualization of any device. There may be two types of printers, one which can be operated manually and the other which can be having inbuilt wi-fi support to manage it from our laptop or phone, so that it can be controlled by the signals. Another aspect of connect is to have high speed messaging, after connecting all devices with the platform and due to generation of a lot of data through these devices, there must be a bi-directional communication between devices and platform. Endpoint management is the major part of connect: to identify the devices from where and to whom the data has to transfer is pretty essential.

Analyze: The real time analysis of data (Harris, 2017) is very important as the data coming is the raw one, so aggregation, filtering, and correlation of data streams is to be done. Once the relevant information is with us we need data enrichment to enrich the data streams with contextual information and generate composite streams. Lastly this enriched data which is surely to be a big data needs to be identified to obtain visualization of that data with integrated cloud service support.

Integrate: This feature of Internet of Things is about enterprise connectivity; this will make an effortless interaction with the service provider to check the problem or call them or to wait for them. Communication with REST API with respect to the cloud applications or IoT will result in more efficient and easy communication. Integration is also about command and control as per users' requirement on the basis of voice-based recognition or by message from their mobile phone.

So above all, features of Internet of Things make this beneficial for society in the form of effortless communication and time saving activity. More light on this will be provided later in this chapter.

B. Benefits of Internet of Things

IoT (Charalabidis, 2019) is expanding the interdependence of people to interact and contribute, which makes it important for the existing technology in various sectors of E-government and society like healthcare, transportation, home appliances, agriculture, industrial applications, military applications, and so on. The major benefits of IoT are as follows:

1 **Efficient resource utilization:** IoT can significantly improve the efficient use of smart devices and advance the harmony between artificial intelligence and environment.

2 **Minimizing human effort and time:** IoT assumes a significant role in reduce human effort. It jumps forward, taking the advantages of most recent remote devices and correspondence technologies. Various applications of IoT are smart homes, smart business, security, and surveillances used for various purposes. IoT has made life simpler for both users and application engineers from various perspectives. With such a great amount of intelligence of the Internet with the devices, innovation has been able to deal with various tasks easily.

3 **Development of AI through IoT:** Artificial intelligence is the technology at which a framework can finish a lot of assignments or learn from data in a manner that appears to be intelligence. In this way, when artificial intelligence consciousness is added to the Internet of Things it implies that those devices can break down information and settle on choices and follow up on that information without contribution by people. Robots and automatic vehicles are examples of IoT. These IoT devices have sensors for gathering the data and take the decision through the AI platform. AI technology trained the model to take appropriate action.

4 **Improved security:** The greatest concern around the Internet of Things (IoT) is ensuring that systems and information and devices associated with them are secure. Therefore, giving IoT security turns into a huge test while protecting devices from destructive assaults and unapproved access.

To sum up the various methodologies, the most important layers that are accessible in most IoT arrangements are described in Figure 4.3.

In Figure 4.3, part A represent the three-layer architecture which is not sufficient for research on IoT. Part B of Figure 4.3 illustrates the five layered architecture. These layers work as given below.

Perception Layer: Responsible for gathering the information and senses some important objects in the environment.

Transportation Layer: Transfers the data from the perception layer through the LAN, NFS, and Bluetooth.

Processing Layer: Analyzes and manages the large amount of data which comes from the transport layer.

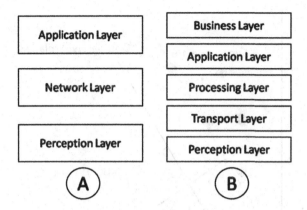

FIGURE 4.3 Sethi and Sarangi (2017). Internet of Things [painting]. Retrieved from www. hindawi.com/journals/jece/2017/9324035

Application Layer: Used for delivering the services to the users. Application layer defines the various applications like smart health devices, smart cities.

Business Layer- It is responsible for manage the whole IoT system.

Network Layer- This layer is responsible for routing of data packets and connecting to network devices and servers.

4.4.2 ARTIFICIAL INTELLIGENCE (AI)

AI is a field of computer science which has the ability to make machine intelligence as human. Artificial intelligence is the advance technology for society and economy implemented at large scale. These developments can possibly straightforwardly impact both the creation and qualities of a wide scope of items and administrations, with significant implication for employment and production. Artificial intelligence makes human life easier and saves time. AI involves the intersection of several advanced areas of other domains including machine learning, natural language processing, computer vision, and many more.

The machine learning algorithm has the ability to learn from prior experience and make the correct decision for the future. Traditional machine learning algorithm is deep learning algorithm which produces the better outcome compare to other ml algorithm. Machine learning algorithms are used in various areas such as medical field, online fraud detection, self-driving car, E-governance, etc.

Medical field research has high value for society. AI is an emerging technique which is used in every field like agriculture, healthcare, banking, and many more. Different numbers of machine learning techniques are as given below:

A. **Support Vector Machine (SVM)** – Support Vector Machine is a supervised type machine learning algorithm. SVM is used for regression and classification, but mostly used for classification problems. This technique produces the lower prediction error as compare to other algorithms. In this technique the

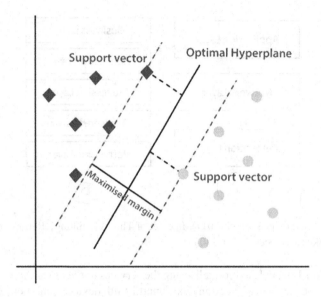

FIGURE 4.4   Java Point. SVM Architecture [painting]. Retrieved from www.javatpoint.com/
machine-learning-support-vector-machine-algorithm

classes are separate by the hyper plane. The prior objective of this algorithm is
to plot the hyper plane that has maximum margin to classify the data points in
to two classes. Maximum margin means maximum distance between the data
points of classes.

The optimal hyper plane is defined as

$$w.x^T + b = 0$$

where w is the weight vector
  x is the input feature
  b is the bias

B. **Random Forest:** Random Forest is also a supervised learning algorithm.
   Random Forest built the model using the multiple decision tree merge them
   together. In this way, Random Forest, just an arbitrary subset of the features is
   taken by this algorithm for parting a node. This algorithm produces the better
   result because it takes the random parameters. Random Forest overcomes the
   problem of decision tree.

C. **Deep Learning:** Deep learning is a very impressive technology inspired by
   the human brain function called the Artificial Neural Network. This algorithm overcomes the problem of other machine learning techniques. A main
   benefit of this algorithm is not requiring feature engineering. The number of
   hidden layers is present between input and output layer. Each layer performs
   the specific type of filtering. First it extracts the features then performs

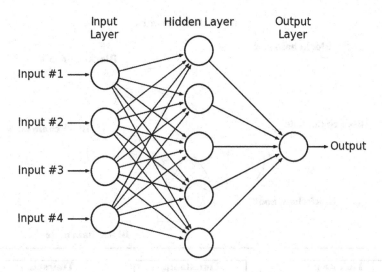

FIGURE 4.5   Luciano Strika (2019). Architecture of Neural Network [painting]. Retrieved from https://www.kdnuggets.com/2019/07/convolutional-neural-networks-python-tutorial-tensorflow-keras.html

classification. In deep learning, different types of neural networks are ANN (Artificial Neural Network), CNN (Convolution Neural Network), and RNN (Recurrent Neural Network). These neural networks are used in the world for different domain like self driving car, speech recognition, diseases detection etc. The neural network works in three layers – input layer, hidden layer, and output layer. Input layer is used for accepting the input, hidden layers are used to process the input, which means extracting the feature, and output layer produces the final outcomes.

ANN works for the text data, CNN works for image classification, and RNN works for natural language processing. Every network is used for different problems.

   D. **k- nearest neighbor (KNN)** – KNN is supervised type learning algorithm. It is mostly used for ease to interpretation and low calculation time. KNN can be used for both classification and regression. K is the number of nearest neighbors. It predict the value of new data points using feature similarity; these new data points will be assigned a value based on how closely it matches the data points in the training set.

4.4.3 BLOCKCHAIN

These days cryptographic money has become a trendy expression both in the industry and in the scholarly world. As one of the best digital currencies, Bitcoin has delighted in colossal accomplishment with its capital market arriving at 1000 billion dollars in

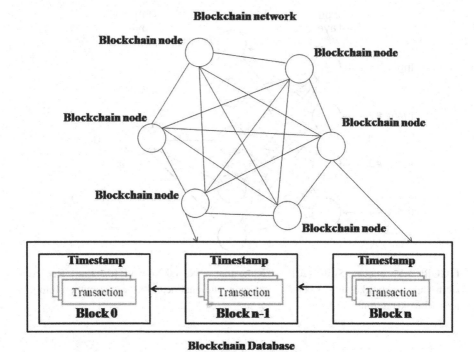

FIGURE 4.6   Salman, Jain, Gupta (2018). Blockchain architecture [painting]. Retrieved from www.researchgate.net/publication/333486562_Probabilistic_Blockchains_A_Blockchain_ Paradigm_for_Collaborative_Decision-Making

2021. With an uncommonly planned information stockpiling structure, exchanges in Bitcoin system could occur with no outsider and the center innovation to manufacture Bitcoin is blockchain.

Since it permits installment to be done with no bank or any go-between, blockchain can be utilized in different budgetary administrations, for example, computerized resources, settlement, and online installment. Furthermore, it can likewise be applied in different fields including savvy contracts, open administrations, Internet of Things (IoT), notoriety frameworks, and security administrations. Those fields favor blockchain in numerous ways. Most importantly, blockchain is unchanging. Exchange can't be altered once it is pressed into the blockchain. Organizations that require high unwavering quality and trustworthiness can utilize blockchain to pull in clients.

Block chain is a grouping of blocks, which holds total transactions of exchange records like customary open record.

This architecture of the blockchain innovation yields many engaging attributes, including appropriated management, decentralized record, trustless accomplices, provable security, changelessness, and non-renouncement ensures. Each term is quickly depicted here. The management is conveyed as the blockchain database is kept up by many "blockchain hubs" and no gathering has full power over the framework.

4.5 AI IN SOCIETY

Artificial intelligence will improve our lifestyle. More efficient tasks will be done by the intelligent system. Some smart devices and smart homes will also reduce the human time and energy and also provide better health care decisions (Arora, 2016.). In the healthcare sector AI can reduce the operating costs and money in operations of healthcare using robotics.

4.5.1 HEALTHCARE

This is bringing a change in perspective to society, fueled by expanding accessibility of human services information and quick advancement of investigation methods. AI can be applied to different type of medical data. Various machine learning technologies such as Support Vector Machine, Deep Neural Network, Naive Bayes, Random Forest are used in healthcare for detection the diseases. Current AI technology (Gudivada, 2018.) has sent tremendous waves across clinical services; an active discussion of AI technology is whether AI technology can replace human specialists or not in future. Be that as it may, AI can help doctors take better decisions or even suggest human decision in certain useful operational areas of biomedical sector.

The expanding accessibility of medicinal services information and quick improvement of big data technique has made conceivable the ongoing fruitful uses of AI in biomedical. Suggested by pertinent clinical inquiries, ground-breaking AI strategies can open clinically applicable data covered up in the huge measure of information, which thus can help medical decisions.

With the help of recent advances AI technology has improved patient care which is low in cost and faster in diagnosis. Deep learning (Raghupathi, 2014) as a neural network is a widely used machine learning technology which provides high accuracy because it builds a model with many hidden layers.

APPLICATION OF AI IN HEALTHCARE:

Drug discovery: AI technology is used for decrease the drug discovery time. Utilizing AI to reestablish portions of the medication revelation procedure can be less expensive, and more secure. Simultaneously AI can't totally evacuate all the stages worried in drug discovery, it can help with stages like, finding new exacerbates that could be potential medications. It can likewise help to discover new applications for recently tested compounds.

Early detection and diagnosis: The patients are suffering for the lack of early detection of disease. Using AI technology builds a model which can detect the sign of the diseases in the early stage so that patients can take timely treatments. Cancer is the second leading problem in the world. The most common types of cancer are lung cancer and breast cancer (Hwang, 2002). Death rates from cancer is increasing day by day. We can prevent it if it is detected in its early stage. Data science technology is useful for this type of detection. Different machine learning algorithms are used to analyze the pattern in data without

Benefits of schools and teachers

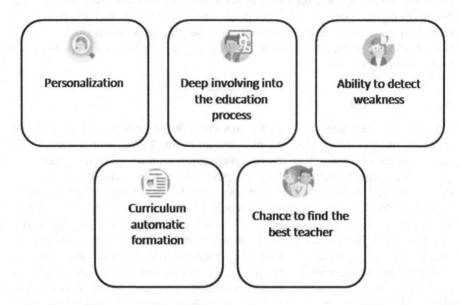

FIGURE 4.7   Kuprenko (2018). Benefits of AI in e-learning [painting]. Retrieved from https://medium.com/towards-artificial-intelligence/artificial-intelligence-in-education-benefits-challenges-and-use-cases-db52d8921f7a

giving any predictions. Artificial Neural Network and support vector machines are an extensively used technology for images diagnosis (Jiang, 2017).

Health monitoring devices: Health tracking devices can monitor the health of the patients and share the information with doctors. These innovation technologies have made human life easier. These devices can detect heart rate and level of blood glucose and can suggest exercise or an additional diet plan.

IoT devices in robotics surgery: AI is applied to surgical robotics (Shaikhina, 2017). Producers see the need to utilize deep learning information to robotize as opposed to conduct customized by a designer who doesn't have a clue about all the situations. This AI data is gathered from watching specialists perform. AI is used with machine learning technology to recognize cancerous cases. Robotics surgery can suggest unexpected missing steps.

4.5.2 EDUCATION

AI technology is useful in make decisions and conclusion. Advanced technology is applied in education to help develop skills. Machine learning technologies provide the platform for teaching with lots of information. AI is becoming part of our normal lives. Today, students don't have to go to physical classes to concentrate as long as they have PCs and web association. Artificial intelligence is likewise permitting the automation of authoritative assignments, permitting institutes to limit the time required to finish tasks so instructors can invest more energy with students.

APPLICATION OF AI IN EDUCATION

Virtual tutorial system: Not to be ignored is the evident fear that human teachers can or will be supplanted by AI advancements in the coming decade. As AI progresses in this space, it appears there is more proof to help the possibility that both intelligent systems and people are expected to oversee various parts of students' scholastic and social capabilities. Simulated intelligence will probably not supplant but will fill in as an important expansion of the human master, helping instructors meet the various needs of many students at the same time all the more successfully. Obviously, only human educators can comprehend researchers' needs better; however it's acceptable to get instant feedback from the virtual guide.

AI based scoring system: Educational institutes are receiving the rewards of AI and ML advances. Criticism and scoring frameworks are planned and created using AI based machine learning algorithms to help students and experts in writing. Grammarly, Turnitin, and PEG Writing are just a few of the ML-based applications created for improving writing skills.

Global learning system: AI provides a global platform by which we can share knowledge all over the world. Students can easily study the various training using AI based solutions. There are a great number of stages with intelligent implementation of materials from the best guides. The innovation is utilized to create portable and online study and test preparation applications, for example, Quizlet, Toppr, etc.

4.5.2 COMMUNICATION

Today, consumers continually communicate with social media. A significant number of us may even say we are dependent on our screens. Organizations are anxious to exploit our consistent commitment with stages, for example, Facebook, Twitter, and Snapchat. Subsequently, a developing number are consolidating artificial intelligence (AI) in online life to all the more likely interface with expected consumers. Effectively, only a solitary snap can affect what warnings spring up on our web based life accounts – posts, promotions, companion proposals, and that's only the tip of the iceberg – as a result of AI items, for example, suggestion motors and chatbots (Zhang, 2018). Moreover, AI innovations, including facial recognition and natural language processing (NLP), are helping organizations improve customer care and market their items all the more adequately.

APPLICATION OF AI IN COMMUNICATION

Online advertising: The general inclination of advertisers is to use the compensation per click publicizing utilizing Facebook or AdWords. Joined with AI, advertisers will have the option to discover more and better promoting stages.

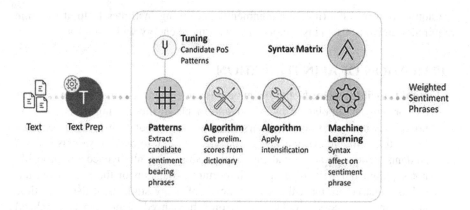

FIGURE 4.8 Sentiment analysis system

Artificial intelligence fueled instruments can be used to investigate, oversee, and streamline paid advertisement crusades. The apparatuses help to discover best channels to convey promotions to your favored audience.

Market Basket Analysis technique is used in the business intelligence world. Market Basket Analysis is a procedure which recognizes the strength of relationship between sets of items purchased together and distinguishes patterns of co-occurrence. A co-occurrence is when at least two things occur together. A priori algorithm is useful in extracting frequent items and relevant association rules. This algorithm is implemented with the database containing a large number of transactions. For example, when a customer buys items from a supermarket then this algorithm helps to buy items easily. This algorithm is beneficial for business and enhances sales.

   Sentiment analysis: The advance technology of AI has the ability to detect the negative words from your tweet. Sentimental analysis uses the Natural Language Processing (NLP) to recognize the negative and positive words in the comment on social media. Sentiment analysis is modeled as a classification problem, where classifiers take a text and extract the words in the category as positive, negative, or neutral. The classification algorithms include the statistical model like Support Vector Machine (SVM), Neural Network, Naive Bayes.

An example of how stock cost of an organization can be influenced by a news occasion, and as you can envision, the assessment communicated in the updates on obtaining can be a trigger for a stock exchanging calculation to purchase the stock before the expansion in its cost occurs.

   Sentiment analysis frameworks combine AI with software rules over the whole content analytics function stack, from low-level tokenization and language structure investigation as far as possible up to the most significant levels of sentiment analysis.

Streamlining communication with chatbots: Chatbots are pieces of programming that direct discussions via auditory or textual techniques and encourage correspondence among buyers and organizations since they are programmed to react promptly to requests, saving time and, generally speaking, improving client experience. Spotify, Wall Street Journal, and Sephora as of now speak with their clients through chatbots on Messenger. Chatbots are a very fast tool for business communication in today's industries.

Chatbots are virtual guides, experts or aides whose undertaking is to converse with an Internet user progressively. In any case, they lead discussions without human intervention. In all actuality, they are computer programs furnished with a special algorithm that empowers discussion and activities identified with your clients' needs. The communication tools are used in customer services, purchase and ordering of products, product consulting, etc.

4.5.3 BANKING

The banking sector is becoming one of the primary adopters of artificial intelligence (Jewandah 2018). Banks are investigating and executing innovation in different manners. Artificial intelligence is showing signs of improvement and becoming smarter day by day. The Indian financial sector is bit by bit moving towards utilizing AI. The Indian financial division is investigating the ways by which it can outfit the intensity of AI to improve the procedures and upgrade customer service.

SBI is the biggest public sector banking tool with 420 million clients having embarked on utilizing AI by launching "Code For Bank" for concentrating on innovations, for example, blockchain (Tara Salman, 2018), machine learning, IoT, AI, and robotics process. SBI has additionally launched SIA, an AI-powered chat that tends to client enquiries in a second and performs regular financial tasks simply like a bank. Descriptive research explored the significance of utilizing artificial intelligence particularly in banks to reduce dependency on humans and likewise to comprehend what can be the potential implications of the utilization of artificial intelligence.

APPLICATION OF AI IN BANKING

Smart wallets: Significant players like Paypal, Google, Apple, and others have developed their own payment gateway. The digital wallet is an easy method of payment in the real world. This advance technology decrease dependency on physical cash, in this way extending the scope of cash to more prominent levels.

Fraud detection: Fraud detection is a major issue in the finance field although it is used in every field like medical, insurance, and many more. It is a very challenging task to keep the system secure and prevent the customer from fraud. Unsupervised and supervised algorithms of machine learning are used in model prediction for detecting unusual transactions. This type of model has the capability to teach itself from past experience and find hidden patterns which help in fraud detection.

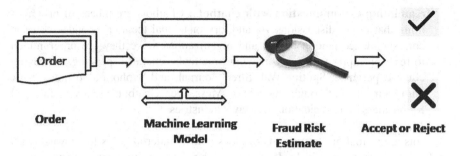

Order   Machine Learning   Fraud Risk   Accept or Reject
                     Model   Estimate

FIGURE 4.9 A priori algorithm in data mining (2020). Fraud detection process model [painting]. Retrieved from www.softwaretestinghelp.com/apriori-algorithm/

Online transactions increase day by day through credit card, debit card, e-wallets, and many more so fraud activities are also increasing. Researcher are trying to solve this major issue using artificial intelligence. Different machine learning algorithms like Decision Trees, Naive Bayes Classification, Least Squares Regression, Logistic Regression, and SVM are used in model prediction which can easily analyze fraud in online transactions.

Here the fraud detection process is first gathering the data in a model and applying machine learning algorithms. Before applying he algorithm, split the data into training and testing. Having trained the model on the specific dataset, the model is able to work for fraud detection which is the fraud risk estimator. The accuracy of the model can be improved when it is applied to more data.

> **Virtual customer support:** As natural processing technology develops, we are moving nearer to the day when machines could deal with most client support inquiries. This would mark an end to hold ups in the process chain and result in more satisfied clients.

4.6 CONCLUSION

The main concern in this chapter has been in improving situations, services, and systems of society which are pillars of a better economy. The most challenging part of a developed country is to build secure, permanent, and transparent human-intensive governance; this is the only point which turns developing countries into developed ones. Thus, the requirement is to go to the next level, utilizing and improving the country's engineering and managing capacity required to design, build, and operate the system of current technologies to ease the complexity of government and society. With recent advances in AI and data science many government firms are trying to upgrade technology to improve their systems and services. This will require lot of good experts, adoption of better technologies, computational resources, and artificial intelligence. Building a new framework for engineering and E-governance will enhance or advance E-governance. Deep learning, IoT, data science, and machine learning will, all together, support in the advancement and development of E-governance. The management of E-government should be end-to-end so that the principles of trust, transparency, and efficiency are fulfilled. The goal can be achieved by coming up with ever more progressive technology in data science and AI.

REFERENCES

Agbozo, E., and Asamoah, B. K. 2019. Data-driven E-government: Exploring the Socio-economic. Retrieved from Ramifications.Je. DEMISSN, 9517, 2075.

Agnihotri, N., and Sharma, A. K. 2015. Big Data Analysis and Its Need for Effective E-governance. *International Journal of Innovations and Advancement in Computer Science*, 6(2). ISSN 2347, 8616.

Gudivada, A., and Tabrizi, N. 2018. *A Literature Review on Machine Learning Based Medical Information Retrieval Systems*. IEEE Symposium Series on Computational Intelligence. SSCI.

Harris, T. H. 2017. *Competing on Analytics: The New Science of Winning*. *Harvard Business School Press*. Harvard Business Review Press.

He, K., Zhang, X., Ren, S., and Sun, J. 2016. Deep residual learning for image recognition. *IEEE Conf. Comput. Vis. Pattern Recognit* (pp. 770–8). IEEE Conf.

https://intellipaat.com/blog/fraud-detection-machine-learning-algorithms/.

https://medium.com/cloohawk/4-applications-of-artificial-intelligence-in-social-media-marketing-745babf7bfe0.

Hwang, K.-B. et al. 2002. Applying machine learning techniques to analysis of gene expression data: Cancer diagnosis. *Methods of Microarray Data Analysis*. NY: Springer, 167–182.

Jewandah, D. S. 2018. How Artificial Intelligence Is Changing the Banking Sector –A Case Study of Top Four Commercial Indian Banks. *International Journal of Management, Technology and Engineering*. 8(7), 525–530.

Jiang, F., Jiang, Y., Zhi, H., Dong, Y., Li, H., Ma, S., . . . Wang, Y. 2017. Artificial Intelligence in Healthcare: Past, Present and Future. *Stroke and Vascular Neurology*, 2(4): 230–43. doi:10.1136/svn-2017-000101.

Kankanhalli, A., Charalabidis, Y., and Mellouli, S. 2019. IoT and AI for Smart Government: A Research Agenda. *Government Information Quarterly*, 36(2): 304–9. doi:10.1016/j.giq.2019.02.003.

Kuprenko, V. 2018. *Artificial Intelligence in Education: Benefits, Challenges, and Use Cases*.https://medium.com/.

Luciano Strika, M. 2019. *Convolutional Neural Networks: A Python Tutorial Using TensorFlow and Keras*. KDnuggets.

McKinsey Global Institute. 2011. *Big Data: The Next Frontier for Innovation, Competition, and Productivity*. McKinsey Global Institute.

Navdeep, P., Arora, M., and Sharma, N. 2016. Role of Big Data Analytics in Analyzing e-Governance Projects. *International Conference on New Trends in Business and Management: An International Perspective*. ISSN 2250-348X.

publicadministration.un.org. 2018. *UN-E-Government-Survey*. en-us: https://public administration.un.org/.

Raghupathi, W., and Raghupathi, V. 2014. Big Data Analytics in Healthcare: Promise and Potential. *Health Information Science and Systems*, 2(3). doi:10.1186/2047-2501-2-3.

Schwartz, L. 2016. Data, Data Science, and the Research University. *IIAI International Congress on Advanced Applied Informatics*. 978-1-4673-8985-3/16.

Shaikhina, T., and Khovanova, N. A. 2017. Handling Limited Datasets with Neural Networks in Medical Applications: A Small-Data Approach. *Artificial Intelligence in Medicine*, 75: 51–63. doi:10.1016/j.artmed.2016.12.003, PubMed: 28363456.

Shreshtha, S., Singh, A., Sahdev, S., Singha, M., and Rajput, S. 2019. A Deep Dissertion of Data Science: Related Issues and its Applications. *IEEE.*

Tara Salman, R. J. 2018. Probabilistic Blockchains: A Blockchain Paradigm for Collaborative Decision-Making. *IEEE UEMCON 2018.* NY.

Venugopalan, S., Xu, H., Donahue, J., Rohrbach, M., Mooney, R., and Saeno, K. 2015, April 30. Translating Videos to Natural Language Using Deep Recurrent Neural Networks. *arXiv:1412.4729 [cs.CV].*

Vijai, C. 2019. Artificial Intelligence in Indian Banking Sector: Challenges and Opportunities. *International Journal of Advanced Research*, 7(4), 1581–7. doi:10.21474/IJAR01/8987.

Zhang, Y.-D., Zhang, Y., Hou, X-.X., Chen, H., and Wang, S-.H. 2018. Seven-layer Deep Neural Network Based on Sparse Autoencoder for Voxelwise Detection of Cerebral Microbleed. *Multimedia Tools and Applications*, pp. 10521–38.

www.softwaretestinghelp.com/apriori-algorithm/.

www.marketingaiinstitute.com/blog/what-is-artificial-intelligence-for-social-media.

www.forbes.com/sites/bernardmarr/2018/07/25/how-is-ai-used-in-education-real-world-examples-of-today-and-a-peek-into-the-future/#27c0fc1b586e.

5 Applying Machine-Learning and Internet of Things in Healthcare

Pushpa Choudhary, Akhilesh Kumar Choudhary, Arun Kumar Singh, and Arjun Singh

CONTENTS

5.1 INTRODUCTION

Medicinal administrations are a principal bit of life. Lamentably, the reliably developing people and the associated climb in constant sickness is setting prominent stress on present day human administrations systems, and enthusiasm for resources from clinic structures to authorities and clinical specialists is exceptionally high. Clearly, a game plan is needed to decrease the weight on human services frameworks and continuing to provide first rate services to the peril patients. The Internet of Things (IoT) has been extensively identified as a capable response to moderate the loads in social protection systems; it is the point of convergence for much late examination. A great deal of this investigation looks at checking patients with specific cases, for instance, diabetes. Additionally research increases to fill certain conditions, for instance, providing support to recuperation with consistent examination of the patient's headway. Crisis social insurance, which has moreover been recognized as a likelihood by associated functions, have for the most part not yet been examined. A few related works have as of late assessed specific territories and headways related to IoT social protection. A broad study is presented, based on financially accessible courses of action, potential applications, and remaining issues. Each point is considered freely, instead of as a major aspect of a general process. In, data mining, amassing-examination is thought of, with minimum notification compromise of them in the process (Aazam et al. 2018).

During the previous decade, the Internet of Things (IoT) has increased in noteworthy consideration in the scholarly community as it has in industry. The primary purpose of IoT capability is what it offers for industry. Nowadays around the world all items are connected to the Internet and speak with one another with the minimum of human intercession. Despite the fact that IoT includes various thoughts and ideas, it doesn't have a reasonable definition. This chapter provides an overview of the meaning of IoT in three alternate points of view: things, Internet, and semantics i.e. "**things**" that sense and collect data and send it to the **Internet**. In the present scenario, IoT devices are an integral part of any network.

Sensor systems are the major empowering influence of the IoT. A sensor can be characterized as a gadget that recognizes or measures a physical marvel, for example, moistness, temperature, and so on. A sensor hub is a physical stage that has at least one sensor. Every sensor hub has the ability to detect, impart, and process information. A run of the mill sensor (Ahamed and Farid 2018) involves at least two sensor hubs which connect with each other utilizing wired and remote methods. In sensor systems, sensors can be homogeneous or heterogeneous. Numerous sensor systems can be associated together through various components. One such methodology is through the Internet. Ordinarily, sensor hubs are sent in thick way around the marvel which we need to detect.

These sensor hubs are minimal effort and though little in size they empower huge organizations. Sensor organize isn't an idea that rose with the IoT. The idea of sensor organize and related research existed long time before the IoT was characterized. This can be plainly observed when we assess the writing in the field. Notwithstanding, the rise of the IoT has encouraged the standard selection of sensor organize as a significant innovation used to understand the IoT vision. As of late, another generally perceived wellspring of sensor information is acquired from versatile keen gadgets. The universal idea of portable brilliant gadgets – for example, advanced mobile phones, tablets, savvy – and the accessibility of modest installed sensors have totally upset the shrewd city application measurements.

5.2   RELATED WORK

In this area we present a determination of late use cases, where gushing information combination has been applied with progress. We separate between basic gushing information combination techniques, which coordinate area information into models, from space rationalist strategies, similar to our own. Gushing information combination is normally stretched out with gradual learning strategies, where we give a fundamental diagram of the best in class. At last, we finish up the area with an introduction of scholarly and creation grade stream handling motors and a review of practically identical gushing information combination stages.

Sensor combination is useful for improving certain functionalities and model exactness in different areas, i.e., in situating and routing, in action acknowledgment in framework observing and issue determination in transport, in social insurance and in others (Ahmadi et al. 2019; Akbar et al. 2017).

In human services, for instance, information combination is utilized in IoT-empowered settings, for example, remote patient checking frameworks. Here, the patient is checked with various body and natural sensors, whose signs are prepared and used to advise the specialists regarding the patient's condition. Information combination is utilized to join the various signals on three unmistakable levels: crude level (combination of crude sensor information), include level (consolidating highlights gave by various techniques), and choice level (consolidating choices or confidences of clinical specialists). Incorporation of our structure in a remote patient checking framework could give extra upgrades (i.e., consideration of authentic information in mix with current qualities). The majority of the referenced sensor combination strategies anticipate that all the information should be intelligible, accessible quickly and showing up in the right request. In many restricted frameworks this is the situation, notwithstanding, in the IoT situations, the accessibility of the information adds to the vast majority of the issues. Access control plays a significant issue in information the board and is a clear theme in ongoing writing (Al-Garadi et al. 2020; Babar et al. 2020).

5.2.1 INTERNET OF THINGS IN HEALTHCARE

Examination in associated sectors exhibited that prosperity watching can be possible, yet progressively noteworthy are points of interest it could give in various settings. Prosperity watching can be utilized to screen non-fundamental patient in house, not in hospitals, decreasing strain on crisis facility, for instance, authorities.

In complete honesty, commonly scarcely any burdens of remote prosperity checking but hindrances add a security risk. On account of various favorable circumstances of remote prosperity watching, various continuous examiners have perceived the ability of IoT as the response for protection. Rebuilding after physical injury has been a subject explicitly critical for a couple of investigators. The structure prepared for recognizing cardiovascular disappointments was amassed using moment sections. Circle is the structure in continuing with progress using common and vision-based for instance camera sensors for basic development and prosperity watching. It thinks about arbitration by means of gatekeepers and pros. Examiners going after the endeavor have perceived that AI would be important for getting some answers concerning situations (Catarinucci et al. 2015; Lee et al. 2014).

Our foundation expands on components, portrayed in the writing and can exploit late discoveries, particularly those identified with gushing stages. Uncommon commitments talk about treatment of postponed or out-of-succession estimations.

A considerable number of the frameworks additionally join huge area information (model-driven methodologies) into the information combination model (primarily into the Kalman channel's progress lattice), by which the models lose their speculation potential. Structures that can possibly be applied in different use cases should be space information freethinker (simply information driven), at any rate with the demonstrating calculation. In this sense, any AI calculation goes about as an information combination model since it joins different markers into a solitary forecast (Candanedo et al. 2018; Hasan et al. 2019).

5.3 SENSOR DATA FUSION

In this section, sensor information combined with the IoT space and its significance towards the IoT is presented. As discussed in previous sections, Internet of Things would deliver a considerable measure of information that are less valuable unless it can determine information utilizing them. The accompanying articulations firmly accentuate the need of sensor information combination and sifting in the IoT area (Verma and Sood 2018).

Information is one of the most important conditions in the IoT situation. Promote clothing to create imaginative administration. In particular, in applications for general urban communities, when distortion detection begins in the range of 50–100 billion DE, it is basic to waves, and naturally and intelligently causes information (Shafiq et al. 20201).

Combination is a wide term than can be deciphered from numerous points of view. Lobby and Llinas have characterized the sensor information combination as a technique for joining sensor information from various sensors to deliver progressively exact, increasingly complete, and progressively trustworthy data that couldn't be conceivable to accomplish through a solitary sensor. Nakamura et al. have characterized information combination dependent on three key activities: corresponding, repetitive, and agreeable. Integral methods putting odds and ends of a huge picture together (Samie et al. 2019). A solitary sensor can't say much regarding the earth as it would be centered on estimating a solitary factor such as temperature. Notwithstanding, when we have information detected through various sensors, we can comprehend the earth in a greatly improved manner. Repetitive implies that equivalent natural factor is detected through various sensors. It assists with expanding the precision of the information. For instance, averaging the temperature esteem detected by two sensors situated in the equivalent physical area would deliver increasingly precise data contrasted with a solitary sensor. It likewise lessens the measure of information that should be dealt with as it joins the two arrangement of information streams together. Agreeable tasks consolidate the sensor information together to create new information (Vallathan et al. 20201).

A white paper distributed via Carnot Institutes has recorded information combination and information separating as two fundamental difficulties for the IoT and its applications, for example, brilliant urban communities. Information combination is an information handling strategy that partners, consolidates, totals, and coordinates information from various sources. It assists with building information about specific occasions and situations which is absurd utilizing singular sensors independently.

Information combination likewise assists with building a setting mindfulness model that assists with comprehension situational setting. The sensor information separating stresses the prerequisite of sifting information to maintain a strategic distance from enormous volumes of information transmission over the system. The most fundamental sensor information combination model that is utilized broadly in advanced mobile phones is an e-compass (Sharma et al. 2018).

It utilizes a mix of 3D magnetometer and the accelerometer to give compass usefulness. Principally, information combination activities can be applied at two levels: cloud level and inside the system level. As shown in Figure 5.1, sensor hubs, brilliant city framework, edge hub, sink hubs, and low level computational gadgets, for example, cell phones have a place with in-organize sensor information preparing. Very good quality computational gadgets – for example, servers – have a place with cloud level handling. The cloud can assist better with understanding the earth by performing complex sensor information combination tasks. Cloud level gadgets approach boundless assets and thus has the ability to apply complex information mining calculations over the information created by enormous number of lower level sensors. In the wake of understanding the Earth, the cloud can produce activities that should be taken fittingly. In-organize sensor information combination is critical to diminish the information transmission cost. As information transmission requires noteworthy measure of vitality, applying repetitive combination activity can lessen the information transmission. In any case, low-level hubs might not have the full perspective on the Earth. In this manner, they will most likely be unable to perform complex activities, for example, agreeable tasks (Mighali et al. 2017; Marculescu et al. 2020).

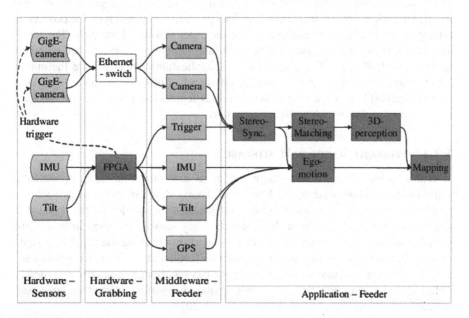

FIGURE 5.1 Sensor data processing.

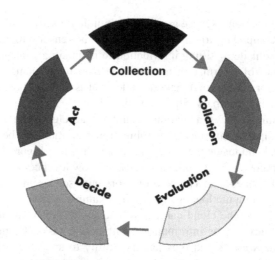

FIGURE 5.2   Internet of Things monitoring cycle.

The principle obligation of in-systems or information combination is to decrease the information transmission cost. The accompanying standard characterizes how the information preparing in each level ought to be led.

The Collation step dissects, thinks about and corresponds with the gathered information. The Evaluation step combines the information so as to comprehend and give a full perspective on nature. The Decide step chooses the activities that should be taken. The Act step essentially applies the activities chose at the past advance. The Act step incorporates actuator control just as sensor adjustment and re-design. Normally, the IoT framework in shrewd urban communities provide a way to screen the ecological setting. There is almost no enthusiasm for the crude sensor information. The information that is of huge intrigue is data about fascinating occasions that are going on in particular territories. To achieve this, IoT applications ought to have the option to catch and reason about the occasions consistently. Subsequently applying sensor information combination strategies at the various degrees of the IoT application chain is basic so as to distinguish pertinent occasions.

5.3.1   WEARABLE SENSOR IN HEALTHCARE

Wearable sensor have been recognized as a key fragment of a therapeutic administrations system set up on Internet of Things development, and as such the progression of careful sensors with low structure factor are major for the productive improvement of such a system. Basically, five key sensors are used in healthcare system. Out of five sensors, three sensors are used for checking the significant signs of heartbeat, respiratory rate, and inner warmth level, and another two sensors are used for watching circulatory strain and blood oxygen, both customarily recorded in a clinical center condition. Some wearable sensors as described below have been used in healthcare (Maddikunta et al. 2020; Shit et al. 2018).

5.3.1.1 Heartbeat Sensors

The heartbeat of a person is the sound of the heart's contraction or the force of blood spreading from one place to another. The number of heartbeats (BPM) per minute is the rate of heartbeat, and heartbeat that can be felt in any artery that is located close to the skin. There are two techniques to measure heartbeat:

- **Manual mode:** Rate of the heartbeat can be measured by examining one of the pulses in two different places; one place may be the neck (carotid pulse) and the other is wrist (radial pulse).
- **Using a sensor:** the rate of the heartbeat can be calculated, which works on optical power based deviation as the beam of light is spread
- or engrossed during its passage throughout the blood as the change of heartbeat.

5.3.1.2 Sensors for Respiration Rate

The quantity of breaths of patient can be observed in each moment. Observation of patient's breath could be in the ID of conditions, as in the following diseases, asthma assaults, hyperventilation because of fits of anxiety, lung malignant growth, blocks of the aviation route, tuberculosis, and that's just the beginning. There are various methods that can be used to utilize the rate of breath.

Echocardiogram (ECG) signs can likewise be utilized to acquire breath rate. This is called ECG Derived Respiration (EDR), and is utilized to decide breath designs and recognize occasions. This strategy peruses respiratory rate sensibly well, although it is again restricted by wear ability. ECG contacts are awkward and would probably disturbe the skin whenever utilized persistently. Moreover, ECG contacts are not reusable and should be consistently supplanted (Mighali et al. 2017; Marculescu et al. 2020).

5.3.1.3 Body Temperature Sensors

The third basic sign is inward warmth level, which can be used to perceive hypothermia, heat stroke, fevers, and anything is possible from that point. Likewise, inner warmth level is a significant diagnostics gadget that should be associated with a wearable human administrations system.

Late works enveloping the estimation of inner warmth level all use thermostat-type sensors. The essential negative-temperature-coefficient (NTC) type temperature sensors were used, while positive-temperature-coefficient (PTC) sensors were thought of. The accuracy of temperature distinguishing is determined by how eagerly the sensor can be put to the human body. In this way, a couple of works focused on making sensors engraved onto humble, versatile polymers with stick backing that could be joined honestly to human skin. While this is an entrancing movement, the work shows that temperature can in like manner be assessed with relative precision using a temperature sensor embedded in materials. Thus, it is recommended that system organizers ought to use materials to hold temperature sensors until equipment engraved on versatile polymer can be even more conveniently created.

5.3.1.4   Circulatory Strain

While not a basic sign itself, circulatory strain is occasionally assessed near to the three vital signs. Hypertension is an acknowledged peril factor for cardiovascular contamination, including respiratory disappointment. It is similarly one of the most generally perceived wearisome sicknesses. Taking everything into account, merging blood pressure into WBAN for social protection can give fundamental data about various patients. We can measure the circulatory strain in the patients with the help of PPT. Endless works have tried to get an accurate measure of BP through estimation of heartbeat travel time (PTT). Some of the work attempted to evaluate PTT property between the ear and wrist, while other work is based on the palm and the fingertip of a hand. PTT is known to be then again relating to systolic circulatory strain (SBP), and is regularly chosen using an electrocardiogram on the chest and a PPG sensor on the ear, wrist, or substitute territory. While no system has yet been created for exactly evaluating circulatory strain diligently utilizing a gently device which is wearable. It is suggested this could be practiced by working up a contraption that utilizes in any event two PPG sensors put along the arm to learn PTT. Circulatory strain is totally a significant parameter in human administrations, and the ability to screen it continually would unbelievably improve the idea of therapeutic administrations that could be given through a WBAN-based structure (Vallathan et al. 2020).

5.4   PROTOTYPE FOR A SYSTEM IN HEALTHCARE USING IOT

In the wake of investigating the wide extent of IoT-based human administrations structure, in this regard wearable sensors are very crucial to each system like remote and remotely wearable sensors. Likewise, fixing or replacing remotely wearable canter points would be essential when compared with inserted vision-based sensors presented in the home. Circulated stockpiling can be used for the purpose of developing prototypes in medical fields.

Circulated stockpiling prepared for taking care of high volumes of fluctuating data was in like manner exhibited to be basic to a significant data human administrations structure by a couple of past works. Using the huge data that will rapidly shape and continue creating in appropriated capacity, these computations could be proposed to mine through the tremendous proportion of data, recognize effectively darkened ailment inclinations. In going with fragments, pieces of the presented prototype are inspected in more depth. Previous prototype is presented and surveyed in the crucial zones. Characteristics and deficiencies of the stream propel are presented, and recommendations for upcoming orientation of examination are also provided.

5.4.1   Central Nodes and Wearable Senor

The focal points of wearable sensors are used for measuring the physical conditions of patient's. The new innovative sensors are basically used for to measure the necessary symptoms like a beat, respiratory rate, and inside warmth level, as these are the central completions of administrative work to confirm fundamental affluence.

After that, sensors can be executed in circulatory stress and blood oxygen, because such constraints are usually in use close to three important signals, for example,

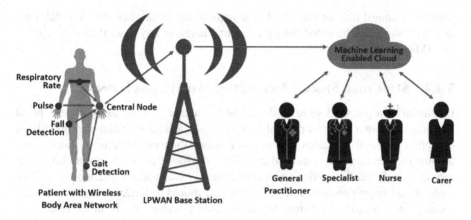

FIGURE 5.3   A process for how wearable sensor works to get information from patient and data transferred to cloud.

specific sensor clarification of the blood-glucose, dip detection, and joint point sensor also focus precisely on the situation that depends on the execution of behavior for structures.

The main central points receive data through the core sensor points that frame this information, possibly accomplishing something unique, and after a short time forward the information to the external field. A pooled central focal point would require a battery backup of the phone which could be improved by pricing appropriate to the Social Security IoT system.

5.4.2   LESS-RANGE COMMUNICATIONS

In order to talk through the central focal point, a short-go special process is required in favor of the sensor. Recalling the effects of the human body, safety, and inaction, there are some requirements to consider when selecting the short-letter correspondence standard. Primary prosperity and rescue vehicles are needed when needed. In those structures, the time delay can be the difference between life and death . In an objective that does not require time; the threshold delay should not be specifically resolved, which is finest at the current time.

5.4.3   WIDE-RANGE COMMUNICATIONS

When chatting using a sensor with a focal focus point, when a specific strategy is needed, recollect the effects of the human body, safety, and sleepiness. The selected strategy for the observation of patient's data, should not have any negative effect on the human body; because it may cause bloating problem for patients. Similarly, concrete security sections should be provided to ensure that delicate patient information cannot be obtained by an antagonist, for a long time, low-latency is basic to time-central modalities, for example, a structure that thrives significantly and requires a rescue vehicle when needed. In such structures, time rearrangement might be the partition amongst demise and life. In the request that is not time-fundamental, low-inactivity shouldn't be sifted through as particularly, in any case, the recent case is best (Maddikunta et al. 2020).

5.4.4 SAFE CLOUD STORAGE ARCHITECTURE AND MACHINE LEARNING

In clinical data procured security should be taken care of when used. Experts profit by understanding a patient's clinical previous cases and AI isn't convincing with the exception of the tremendous databases of information available to it. Taking into account the composition, circulated capacity is the most achievable system for taking care of data. Regardless, offering receptiveness to human administrations specialists without exchanging security is a key worry that should be tended to by pros making social protection IoT structures. AI can be used to ensure safety in the healthcare systems. They help in training the machine so that the machine can do the work done by the people; that ultimately increase privacy as well as the efficiency of the work.

Besides, AI has constantly been perceived in the composition as a strategy for improving restorative administration systems and anyway, it has not been for the most part examined. Man-made intelligence offers the likelihood to perceive floats in clinical data that were at that point cloud based, providing conduct strategy for best diagnostics, and offering suggestions to restorative administration specialists with the purpose of being expressed to particular patients. Taking everything into account, dispersed capacity models should be proposed to help the utilization of AI on gigantic enlightening assortments. The regular model projected for coordinating headway of future Internet of Things (IoT) human administration structures have different use cases. To present the situation, this subpart looks at a couple of those use cases that join sustaining recuperation serving the leading body of unremitting circumstances (Ahamed and Farid 2018).

In addition, it can be used to build up a system equipped to assist with the institute of continuous circumstances, for example, hypertension.

Strains operating between parts of the body over a period of days can be examined and methods are provided by the cloud for the central point worn by the wrist. Yet again, a thorough record of the patient's heartbeat can be the record and the AI system can be used to visualize blueprint for example, while the patient's communication strain is usually considerable.

Patients' information can be used to allow the patient to decide the correct measures to manage their events in order to allow any solution to be used, and the patient must place the bell or caution at the central focal point. Also, help remind the patient to use it.

For example, Parkinson's disease is a model of change in people with dynamic conditions according to which it can be investigated using a systemic system.

The signs of Parkinson's disease strengthening have led to actions, failures, settling issues, and balance issues. With the use of a movement of a wearable accelerometer, sensors can be built into the network to check all these obstacles.

Parametric records can be reliably taken at a set of data that breaks and sent to the central point worn by the wrist, which subsequently forward that data to the cloud in real-time. As data collection from the patient, AI can be analyzed and used to understand the rate at which specific parameters are decreasing for the patient.

Finally in AI, the fundamental technological richness can be seen with an innovative structure in which sensors can be worn, showing real and other huge signals in real-time, including respiration rate, internal heat levels, communication stresses, and pulsation, which are included records that can be taken as a normal, and if any of these parameters fall below the threshold value, the central focal point can spread an alert signal of information in the cloud, and that signal used to instruct urgent situation for the institutions.

In an emergency, real-time readings can be recorded for patient's future treatment and recorded prosperity data available in the cloud, and based on the patient's data which is available on the cloud, specialists can analyze and report treatment based on their reports. As the numbers of persons who regularly undergo diagnosis experience any ill effects in emergency prosperity situations, they have to add new information to their records, using AI to make connections between signals and potential discoveries. After that, this sensitive information can be passed on to paramedics, who can ensure that patients get proper treatment for their existing condition, and increasingly experts try to work hard on this particular treatment system so that they can follow this system in their future work (Verma and Sood 2018).

5.5 APPLICATION AREA OF IOT IN HEALTHCARE

In the present scenario, the number of IoT devices has been applied in the healthcare system. Healthcare sector always seeks new technology and approaches for providing services to patient which reduce delivery time of medicine or things related to healthcare, and also to reducing the costs and improving the healthcare quality. Some of the main areas where IoT has been used in the healthcare system are listed below:

5.5.1 HOME HEALTHCARE

It has been observed by the World Health Organization (WHO) report on aging and disabilities, life expectancy of people has been improved and it is expected that most people globally will live beyond the age of 60. Aging people are more susceptible to chronic diseases, disabilities, and higher hospitalization. Due to this, the researcher's future expectation in the healthcare field they are trying to transfer services related to hospital at home. Home healthcare services based on IoT will reduce the number of things like crowd in the hospitals, hygiene problem, and aging factors affecting ability to go to hospitals frequently and difficulties associated with population aging. In this context, several sensors have been used to monitoring patient condition. Integrating various IoT components into home care and medical systems has become increasingly popular, mainly for events such as fall detection and seizure detection.

5.5.2 MOBILE HEALTH AND ELECTRONIC HEALTH

The development of communication devices such as smartphones and their integration with various types of sensors highlight the increasing usage of IoT technologies. Developed sensor has been used to collects physiological signals of the human body by the software application which is installed in the smartphone. These captured signals are then securely transferred to healthcare institutions. Depending on the patient situation, these signals are converted into short messages which are used to notify healthcare professionals about medical emergency of patient to hospitals and assist them in taking suitable actions.

5.5.3   HOSPITAL MANAGEMENT

In hospital management, responsibilities related to prevention of hospital infections, determining a complete plan for educating patients, management of emergency situations, and logistics systems have been managed. Nowadays we can observe that most hospitals have applied IoT-based technologies such as sensors, ZigBee, RFID, and NFC, and can offer valuable solutions to overcome barriers in hospital logistic management. Even patient can take online appointment for treatment, reducing waiting time of patient. This revolution can enhance hospital management system very easily.

5.6   CONCLUSION AND FUTURE SCOPE

The activities learned through coordinating this outline highlight a couple of areas for extra investigation. To the extent of corresponding measures, it is valuable to make wearable restorative administrations systems. As this is an exceptionally new guideline, no acknowledged prototype has executed restorative administrations condition notwithstanding its prominent inclinations in this zone. Man-made intelligence is a different piece of data to be set up that can be incredibly critical in social protection circumstances. At whatever point applied in the amazing enlisting state for the cloud, AI can offer diagnostics to patients and make new revelations about disorder.

In future, data amassing using cloud developments is to be comprehensively studied, yet data management is the sector in which more upcoming research should be coordinated. Estimations of cloud based data can be done with rough data and concentrate significant data. Hence it has a great deal of chance to show signs of enhancement in privacy and insurance for human administrations which are cloud based. Not acknowledged is that encryption plotting is best for tying down information while offering receptiveness to affirmed social events and engaging AI.

REFERENCES

Aazam, M., Zeadally, S. and Harras, K. A. 2018. Offloading in Fog Computing for IoT: Review, Enabling Technologies, and Research Opportunities. *Future Generation Computer Systems*, 87: 278–89.

Ahamed, F. and Farid, F. 2018. December. Applying Internet of Things and Machine-Learning for Personalized Healthcare: Issues and Challenges. In *2018 International Conference on Machine Learning and Data Engineering (iCMLDE)* (pp. 19–21). IEEE.

Ahmadi, H., Arji, G., Shahmoradi, L., Safdari, R., Nilashi, M., and Alizadeh, M. 2019. The Application of Internet of Things in Healthcare: A Systematic Literature Review and Classification. *Universal Access in the Information Society*, pp. 1–33.

Akbar, A., Khan, A., Carrez, F., and Moessner, K. 2017. Predictive Analytics for Complex IoT Data Streams. *IEEE Internet of Things Journal*, 4(5): 1571–82.

Al-Garadi, M. A., Mohamed, A., Al-Ali, A., Du, X., Ali, I., and Guizani, M. 2020. A Survey of Machine and Deep Learning Methods for Internet of Things (IoT) Security. *IEEE Communications Surveys and Tutorials* 22(3):1646–85.

Babar, M., Tariq, M. U., and Jan, M. A. 2020. Secure and Resilient Demand Side Management Engine Using Machine Learning for IoT-Enabled Smart Grid. *Sustainable Cities and Society*, 62: 102370.

Candanedo, I. S., Nieves, E. H., González, S. R., Martín, M. T. S., and Briones, A. G. 2018, August. Machine Learning Predictive Model for Industry 4.0. In *International Conference on Knowledge Management in Organizations* (pp. 501–510). Springer, Cham.

Catarinucci, L., De Donno, D., Mainetti, L., Palano, L., Patrono, L., Stefanizzi, M. L. and Tarricone, L. 2015. An IoT-Aware Architecture for Smart Healthcare Systems. *IEEE Internet of Things Journal*, 2(6): 515–26.

Hasan, M., Islam, M. M., Zarif, M. I. I., and Hashem, M. M. A. 2019. Attack and Anomaly Detection in IoT Sensors in IoT Sites Using Machine Learning Approaches. *Internet of Things*, 7: 100059.

Lee, B. M. and Ouyang, J. 2014. Intelligent Healthcare Service by Using Collaborations between IoT Personal Health Devices. *International Journal of Bio-Science and Bio-Technology*, 6(1): 155–64.

Maddikunta, P. K. R., Srivastava, G., Gadekallu, T. R., Deepa, N., and Boopathy, P. 2020. Predictive Model for Battery Life in IoT Networks. *IET Intelligent Transport Systems*, 14(11): 1388–95.

Marculescu, R., Marculescu, D., and Ogras, U. 2020. August. Edge AI: Systems Design and ML for IoT Data Analytics. In *Proceedings of the 26th ACM SIGKDD International Conference on Knowledge Discovery and Data Mining* (pp. 3565–6).

Mighali, V., Patrono, L., Stefanizzi, M. L., Rodrigues, J. J., and Solic, P. 2017. July. A Smart Remote Elderly Monitoring System Based on IoT Technologies. In *2017 Ninth International Conference on Ubiquitous and Future Networks (ICUFN)* (pp. 43–8). IEEE.

Samie, F., Bauer, L., and Henkel, J. 2019. From Cloud Down to Things: An Overview of Machine Learning in Internet of Things. *IEEE Internet of Things Journal*, 6(3): 4921–34.

Shafiq, M., Tian, Z., Bashir, A. K., Du, X., and Guizani, M. 2020. Corrauc: A Malicious Bot-IoT Traffic Detection Method in IoT Network Using Machine Learning Techniques. *IEEE Internet of Things Journal*, 8(5): 3242–54.

Sharma, S., Chen, K., and Sheth, A. 2018. Toward Practical Privacy-Preserving Analytics for IoT and Cloud-Based Healthcare Systems. *IEEE Internet Computing*, 22(2): 42–51.

Shit, R. C., Sharma, S., Puthal, D., and Zomaya, A. Y. 2018. Location of Things (LoT): A Review and Taxonomy of Sensors Localization in IoT Infrastructure. *IEEE Communications Surveys and Tutorials*, 20(3): 2028–61.

Vallathan, G., John, A., Thirumalai, C., Mohan, S., Srivastava, G., and Lin, J. C. W. 2021. Suspicious Activity Detection Using Deep Learning in Secure Assisted Living IoT Environments. *The Journal of Supercomputing*, 77: 3242–60.

Verma, P. and Sood, S. K. 2018. Cloud-centric IoT Based Disease Diagnosis Healthcare Framework. *Journal of Parallel and Distributed Computing*, 116: 27–38.

6 Artificial Intelligence
The New Expert in Medical Treatment

Aman Dureja, Aditi, and Payal Pahwa

CONTENTS

6.1 INTRODUCTION

The constant development of medical understanding has complicated this for doctors to stay updated outside their small field. Consulting with an expert doctor Is a remedy whenever the issues are clinical so far beyond the expertise of the healthcare professional, but the analyst's advice is sometimes incomplete or late. Endeavors have been undertaken to build up software programs which can serve as powerful tools (Schwartz 1970; Kissick 1969; Feinstein 1973). It became clear throughout the early 1970s that statistical techniques, such as data flow diagrams, pattern classification, and Bayes theorem, could not address the most challenging medical conditions (Gobry 1973). Researchers then started to research with a professional physician for comprehensive knowledge on the fundamental essence of health problems (Kassirer 1978; Elstein 2000; Michael Chung, Johnson, and Todd 1997; Kuipers and Kassirer 1984). The findings of these works have established the foundation for the theoretical models of cognitive processes, and certain models have since been turned into structures called artificial intelligence (Pauker, Kassirer, Gorry, and Schwartz 1976; Shortliffe 1976; Weiss, Kulikowski, Amarel, and Safir 1978; Miller, Pople, and Myers 1982).

Many of the first initiatives to implement artificial intelligence approaches to specific issues, especially medical imaging, were largely focused on rule-driven systems (Duda and Shortliffe 1983). Such projects are commonly simple to make, in light of the fact that their data is written as "if….do" rules. This is valid for a few little clinical fields, but the most genuine clinical issues are so broad and complex that immediate endeavors to assemble huge arrangements of rules face incredible trouble. Issues emerge principally from the way that frameworks intended to preclude don't mirror the infection model or clinical reasoning. Without such models, the expansion of new guidelines prompts a surprising connection between the principles and accordingly to a huge disintegration of program execution. The reason here is to give an investigation of man-made brainpower to a doctor who has had little communication with software engineering. This won't center around singular plans; rather, this will draw on the basic acknowledgment of such frameworks to make a useful image of clinical computerized reasoning and promising bearings as the field pushes ahead. An AI clinical counsel ought to have a clinical data store that is introduced as depictions of potential illnesses. Based on the nature of the clinical area, the amount of theories in the sample can range from a few to a hundred thousand. In the least complex case of such data, the conclusion of every ailment distinguishes every conceivable part of a specific issue. In fact, the device will be in a position to balance what is understood about the individual and its storing of knowledge.

The simplest version of these structures functions as follows where a significant allegation is raised and when relevant details are consequently received.

1. Determine whether the findings are expected to be appropriate for each disease (diagnosis).
2. Provide a ranking (diagnosis) for each illness in the database by estimating the estimated number of treatments.
3. Classify possible diseases (diagnosis) as per their ailments.

The power of this kind of simplified situation can be greatly improved using a quiz device designed to provide valuable information. Extending the basic plan with the preceding plan, for example:

4. Pick a good high-quality hypothesis and request if one of the natures of the disease can be regarded present or not?
5. If the results do not match with any disease in database, start again with step 1; otherwise, the best possible hypothesis.

Man-made consciousness encourages social insurance groups to downsize documentation time by putting away patient's information carefully and structure a computerized database, which may additionally be utilized for conclusion, treatment and customary Medicare upheld definitive prerequisites, clinical masters in discussion with programming and equipment experts need to build up a stage for information assortment and routine assignments. Non-exclusive programming is being tweaked for explicit applications. In the current situation, AI appears the least complex innovation that might be utilized for giving higher life expectancy. It gives AI-helped automated medical procedures to a modern case. This innovation makes data through various virtual help and every now and again speaks with the patient. In provincial zones, there's a deficiency of social insurance suppliers and this innovation are frequently beneficially utilized to satisfy this lack. It improves the standard of clinical understudies to satisfy any dire interest for provincial zones. This innovation improves the productivity of wellbeing experts as well as improves the standard of human services administration at a lesser expense. It gives guidance to specialists towards the exact conclusion.

6.1.1 BENEFITS OF AI IN MEDICAL FIELD

Computer-based intelligence can tackle distinctive clinical difficulties, such as looking at a unique degree of troubles while producing complex medical procedures with better quality and result. Presently the patient can appreciate the ideal and exact choice. The different benefits of AI inside the clinical field regions:

1. Predict future diseases
2. Improve medical outcomes
3. Accurate, efficient, and quick diagnosis
4. Improve safety in hospital
5. Supportive for complicated and modern care
6. Proper monitoring of patient
7. Checking anomalies and suggesting medical intervention
8. Maintain clinical record
9. Offer support to physicians and patients
10. Good preparation of medical graduates
11. Gather data during treatment to further enhance the longer-term procedure
12. Boost the medical/surgical interaction
13. Improved clinical effect

6.2 COMPUTER AIDED DIAGNOSIS

PC supported determination (CAD) has become business as usual work of the facility for diagnosing malignant growth at mammograms in many testing places in America (Freer and Ulissey 2001; Gur et al. 2004; Birdwell, Bandodkar, and Ikeda 2005; Cupples, Cunningham, and Reynolds 2005; Morton, Whaley, Brandt, and Amrami 2006; Dean and Ilvento 2006; Destounis et al. 2004; Gilbert et al. 2006). It is conceivable that CAD will be broadly utilized for the discovery and conclusion of numerous types of an anomaly in clinical pictures through different imaging strategies, including atomic medication. Albeit early endeavors were made in PC picture examination during the 1960s, broad CAD research started in the mid-1980s at the Hurt Rossman. Radiologic Image Research Laboratories in the University of Chicago's Department of Radiology, with an essential move in the utilization of PC imaging, from robotized PC diagnostics to PC helped diagnostics (Doi 2003, 2004, 2005, 2006, 2007; Lodwick, Haun, Smith, Keller, and Robertson 1963). Computer-aided design PC checking is utilized as a "second conclusion" to help decipher the radiologist's picture. The automated CAD result is utilized as a "second supposition" in helping with the understanding of radiologist pictures. The PC calculation typically comprises of a few stages, which can incorporate picture handling, picture examination, and information preparing utilizing apparatuses, for example, the neural information organizes (ANN); this can be called man-made reasoning.

6.2.1 Detection of Interval Changes in Whole Body Bone Scan

To utilize CAD frameworks that work in clinical circumstances, the radiologist must be sure and feel good with the CAD surge. One powerful strategy utilized by the CAD program to fulfil this necessity is to copy the radiologist's techniques for deciphering clinical pictures. At the point when a patient has a past picture accessible, the radiologist ordinarily looks at the grouping of two pictures to identify resting changes. In this examination study, radiologists utilize the past picture as a veiled picture to improve recognize any adjustments in the current picture. The fleeting deduction strategy discharged for the computerization of the radiologist's concealing procedure, bringing about a better dynamic and decreased CAD learning time.

Improvement of the transient deduction approach was begun with chest radiographs (Kano, Doi, MacMahon, Hassell, and Giger 1994). The clinical utilization of a transitory expulsion strategy is best when the equivalent radiological assessment is performed more than once so that there is a more prominent possibility that patients will get in excess of two continuous interpretive pictures. Accordingly, a chest radiograph was generally fitting to present a fleeting deduction strategy. Notwithstanding, because the patient's definite state of chest radiograph differs to some degree, the straightforward deduction technique couldn't be utilized to the fleeting deduction. Along these lines, a steady picture correlation process has been produced for two continuous pictures. Figure 6.1 shows a nonlinear picture distorting strategy. To this end, a two-dimensional polynomial twisting region was found with a change estimation of x, y facilitates dependent on the number of cross-joins in every one of the relating little locales (ROI) (Kano, Doi, MacMahon, Hassell, and Giger 1994). After the principal endeavor at the expulsion of the chest with radiology was effectively finished and the help of this technique was shown by the investigation spectator, an improvement was found in the decrease of antiquities of inappropriate enlistment.

Figure 6.2 delineates a general automated plan of a transient deduction technique for the recognition of span changes on progressive entire body bone sweep sets. For utilization of a nonlinear picture twisting procedure (Li, Katsuragawa, and Doi 2000), to the transient deduction technique, the grayscale of each picture was standardized first, at that point the measurements, direction, and grayscale of a past picture was changed in accordance with coordinate those of a current picture.

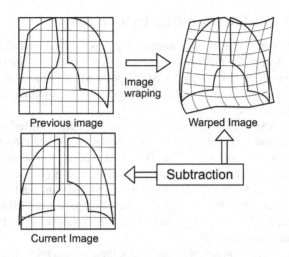

FIGURE 6.1   Visualization of nonrigid warping of image.

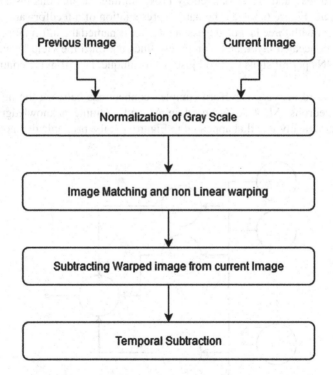

FIGURE 6.2   Computerized workflow for temporal subtraction.

6.3   ANN IN COMPUTER AIDED DIAGNOSIS

An Artificial Neural Network is a nonlinear computational model that copies an organic neural system for the handling of data. It's one of the chief valuables and Efficient methods of computing in AI. At the point when utilized suitably, an ANN can take care of exceptionally complex issues and produce astounding outcomes that customary methodologies can't.

Accordingly, ANNs are utilized as often as possible in different applications, including clinical imaging. for example, analysts have utilized ANNs to identify and analyze interstitial lung illnesses (Wu, Doi, Giger, Metz, and Zhang 1994; Asada et al. 1990; Ashizawa et al. 1999), lung malignant growth (Nakamura et al. 2000; Matsuki et al. 2002; Aoyama, Li, Katsuragawa, MacMahon, and Doi 2002), carcinoma (Zhang et al. 1994; Wu et al. 1993; Baker, Kornguth, Lo, Williford, and Floyd 1995; Lo, Baker, Kornguth, Iglehart, and Floyd, 1997; Sahiner et al. 1996), prostatic adenocarcinoma, and embolism.

An ANN normally comprises of an outsized number of exceptionally interconnected counterfeit neurons (preparing components) working in corresponding to disentangle explicit issues. In most such applications, neurons are associated during a feed-forward way and are actualized by programming calculations instead of electronic gadgets. Figure 6.3 is a schematic representation of a feedforward counterfeit neural system with three layers; the essential layer is named the information layer, the second the concealed layer, and along these lines the third the yield layer. The feed forward ANNs permit signs to travel just in a technique, i.e., from the contribution to the yield layer.

Feed-forward network can learn complex connections between the info neurons and yield neurons. Most ANNs used in AI and example acknowledgment have structures practically like that appeared in Figure 6.3; the principle distinction is that

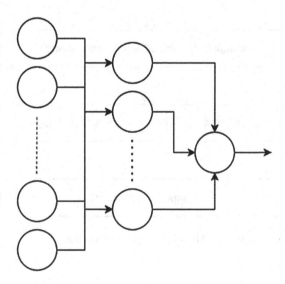

FIGURE 6.3   Feed forward ANN with three layers.

the number of neurons in each layer. for instance, in the event that one intends to utilize three highlights (breadth, difference, and level of circularity) to arrange lung knobs into two classifications what he/she should do first is to build an ANN with three info neurons to point the three-element esteems and one yield neuron to point the probability of being harmful. He/she at that point must train the built ANN by utilizing numerous lung knobs with known highlights for the information neurons and knob classes for the yield neurons. After proper preparation, the ANN ought to be prepared to gain proficiency with the mind-boggling connection between the highlights and in this way the classes and anticipate the classification of a substitution knob by utilizing its highlights. An ANN was first used in PC helped to find for clinical analysis of interstitial lung maladies in 1990 (Wu, Doi, Giger, Metz, and Zhang 1994). From that point forward, ANN has been generally used in CAD plans for recognition and conclusion of fluctuated maladies in a few imaging modalities. during this segment, we represent considerable authority in a run of the mill utilization of ANN in CAD plans, i.e., the clinical analysis of lung knobs and interstitial lung ailments in chest radiography and mechanized tomography.

6.3.1 FUNDAMENTAL STEPS IN DIAGNOSIS

The work process of ANN investigation emerging from the delineated clinical circumstances appears in Figure 6.4 which gives a fast review of the essential advances that should be followed to utilize ANNs for the requirements of finding with adequate certainty. The program gets patient knowledge to forecast the identification of a certain disease for the explanations talked of. Once the empirical condition is set up, choosing the highlights (e.g. side effects, study base, and functional information) to provide the information needed to segregate the patient's various health states should guarantee improvement. This will be cleared out in different ways. Instruments used in chemo measurements permit the end of things that give just excess data or individuals who contribute just to the clamor. In this manner, a cautious choice of reasonable highlights must be regulated inside the primary stage. In the subsequent stage, the database is made, approved, and "cleaned" of anomalies. In the wake of preparing and confirmation, the system is frequently used by and by to foresee the determination. At last, the foreseen conclusion is assessed by a clinical expert. the fundamental advances are regularly summed up as:

1. Selecting important features
2. Generating the database
3. Use of ANN trained on the database to verify
4. Testing in practice

6.3.2 CARDIOVASCULAR DISEASES

These are classified as those disorders impacting all arteries and veins in the guts or blood vessels. They're one of the countries' leading causes of death. In line with the National Health Statistics Centre, CVD is the leading explanation of death within us. Thus, in the last 20 years, CVD has become a crucial field of study. Supporting

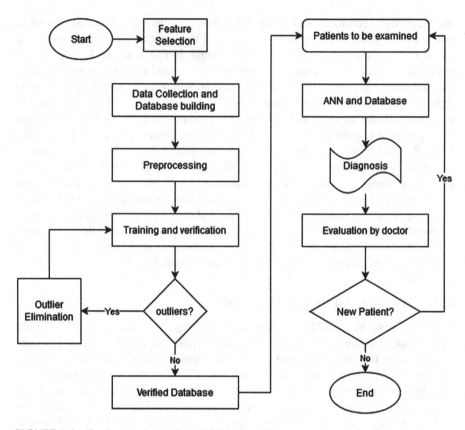

FIGURE 6.4 Fundamental steps in Artificial Neural Network.

a search by book, as many as 1,000 papers have been published since 2008 on usage of ANNs in there and related to cardiac disorder subjects. Arteria coronary disease is right now the world's new leader in definition of mortality in line with the NCHS, Sooner therapy is therefore vitally important. To this end, Karabulut and Ibrikçi implemented ANNs with a back-propagation algorithm as the rotational forest ensemble method's base classifiers (Karabulut and İbrikçi 2011). CAD diagnosis with 91 percent accuracy was obtained from non-invasively, cheaply, and clearly patient-collected data.

Certain details such as height, cholesterol of different kinds, or arterial hypertension are widely used to diagnose CAD (Atkov et al. 2012). The model that emerged with the highest precision (93 percent) was the one that used the disease-related genetic and non-genetic factors.

Notwithstanding these positive findings, it should be remembered that the precision was less than 90 percent for a few examples. ANNs has also extended to other heart diseases, such as abnormalities in the heart valve (Uğuz 2010). Recordings of Heart sound were collected from 120 participants and categorized by stethoscopy as pulmonary, normal, and mitral valve stenosis of heart valve diseases. For 95 percent of the different tones, the right interpretation was accomplished. Two separate programs formulated by Özbay for the detection of arrhythmias obtained a mean accuracy of 99.8 percent and 99.2 percent (Özbay 2008). Consequently, ANNs are also applied to the medical treatment of completely specific conditions, such as the identification of arrhythmias or coronary heart diseases, which are significant causes of death around the world. Usually, classification accuracies of 90 percent are obtained, and in some cases approaching 99 percent. Consequently, ANNs have important promise in the treatment of CVDs.

6.3.3 Diabetes

In developing countries, diabetes poses a major ill health, with projected figures exceeding 366 million cases of diabetes worldwide in 2030 (Leon, Alanis, Sanchez, Ornelas-Tellez, and Ruiz-Velazquez 2012). Type II is the most prevalent form of diabetes, where the immune responses to insulin are compromised and cellular homeostasis and hyperglycemia are disrupted. Direct measures of glucose content in blood samples are the standard of diabetes diagnosis or control. Non-invasive methods for monitoring glucose levels assisted by near infrared or Raman spectroscopy were developed in the early 1980s and are also now usable as a mobile application (Arnold 1996). The networks generalize concentrations of glucose from the spectral curve, allowing for easy diabetes control during day-to-day activities. Standards of living itself has been described as an integral outcome factor in DM management, namely happiness, social experiences and sadness. Narasingarao et al. recently built a test model of a neural network to live the living standard in diabetic patients (Narasinga Rao, Sridhar, Madhu, and Rao 2010).

Evolutionary or pictorial variables such as age, class, weight, or fasting plasma glucose were used as input data for that specific reason. Comparable findings are those from conventional mathematical approaches. In 1997, modeling experiments were conducted on neural predictive regulation of concentration of glucose in tissues (Trajanoski, Regittnig, and Wach 1998).

Trajanoski and collaborators merged ANNs and nonlinear statistical monitoring systems as an effect approach. This strategy has provided for massive levels of noise and time delays. Nonetheless, insulin or analogs administration was found to be inadequate for quick monitoring and was necessary enough to manage sluggish disturbances. A substitution method called a neuro-fuzzy idea was initially developed that uses functional logic rules and neural networks that require a minimum number of input files for proper output (Dazzi et al. 2001). This technology is extremely available for rapid command of intravenous diets in gravely ill people with diabetes and decreased the need for excessive blood sugar evaluation and even the risk of hypoglycemia. According to a beforehand study in patients with type I diabetes, blood sugar levels were also tracked using optimal reverse neural control (Leon, Alanis, Sanchez, Ornelas-Tellez, and Ruiz-Velazquez 2012). RNN used to regulate the high insulin levels following a reference path (a healthy person's normal glucose absorption) to avoid hypoglycemia and hyperglycemia. ANNs have been used not only to glucose monitoring as well as in disease diagnosis. A diagnostic system based on the neural network was developed (Chan, Ling, Dillon, and Nguyen 2011) with the purpose of determining fatal periods of hypoglycemia in patients with type I diabetes. Data from even a cohort of 420 patients was obtained including:

1. Bodily conditions,
2. Pulse-change pace,
3. Corrected electrocardiogram pulse QT length, and
4. Level of shift in the QT interval corrected.

Information from 76 percent patients used to train the neural network, and then the others for testing. A 79.30 percent responsiveness and 60.53 percent accuracy were reached, with greater score than those reported using other techniques such as linear regression methods. Finally, the use of Artificial Neural Network in the onset of diabetes improves the sensitivity, specificity, and accuracy compared to other methods, thereby contributing to better medical practice of DM.

6.4   NLP AND CDSS

There are higher expectations encircling possible improvements endorsed by boosted NLP application in healthcare provision, efficiency and efficacy. For example, one requirement is to incorporate NLP-based "syndromic analysis" of the data on public health. The CDC explains syndromic coverage as the result of the terrorist attacks of 9/11 and subsequent attacks on anthrax ("What is Syndromic Surveillance? on JSTOR" 2020). The goal of symptomatic monitoring was initially to get relatively early, sentinel notice of strategic partners, regional, or national health challenges from various data streams in near real-time, rather than anticipating healthcare organizations to report retrospectively and reactively. Several additional terror group probable benefits have been identified since the first surge of interest in 2001, along with unanticipated transmission of disease outbreaks such as severe acute respiratory syndrome (SARS), avian flu, Ebola, and even Zika. Families, relatives, and individual people usually use unofficial identifiers to explain and classify their diseases. In addition, few will ever dream about using systematic descriptive terms within the ICD-10 scheme. Typically, laypeople would describe their signs, diseases or psychiatric disorders with phrases or terms such as "bones hurt," "tummy-ache," "sleepless," or "on fire." Those words are imprecise though loosely concise, and they often lack meaning. If the tummy-damage happens directly after consuming year-old left-over, the likely sense is very different from a young adult who just stocked up on free soda or even a traveler who just came back from an African Ebola-stricken region! These findings are most also compiled and evaluated in combination with related studies that foreign organizations have generated.

The Clinical Decision Support System (CDSS) is an artificial intelligence based software device that can help doctors, nurses, physicians, or other clinicians make informed choices. Standard use of CDSS is for researching past, existing, and fresh client records and identifying or suggesting user gaps that exist, inaccuracies, safety issues or enhancements in the patient journey. CDSS is also assisted by either of the two types of AI: ES or ML. Most EHR programs provide advanced rule compliance components to incorporate the primary type of CDSS, ESbased CDSS. Many clinicians may specialize in different forms of medical operations, such as lung transplant, and may formulate personalized "order packs" that effectively instigate most of the patient's treatment to ensure optimum outcomes. For example, the order package may include presurgical laboratory testing, prophylactic medicine and instructions of patients, personalized preparedness of trays and sets of medical tools, prostheses, sutures and essential products, video enhancement structures and surgical equipment, postsurgical patient recovery, healthcare, chronic pain medication, psychotherapy, doctors office discharge, and, finally, ambulatory and personal care processes could components of the collection of instructions can provide alternate or alternate measures that impact the changes and circumstances of planned treatment. ML-based CDSS may also be backed by making analyses such as "When the clinical and medical circumstances of this patient are much like the majority of other patients with a similar type of illness, then A, B, and C would generally be the most appropriate diagnosis." IBM's Watson Health is one of those kinds of illustrations. IBM groups with well-regarded healthcare teams, such as Remembrance Sloan Kettering Cancer Hospital or Cleveland Heart Clinic or Orthopedic Procedure. IBM and therefore the physicians "train" an instance (i.e., a fanatical Watson Health System) by using the health history of that clinic making a qualified CDSS tool that will readily match the integrated practices of that hospital (Malin 2013). Nevertheless, the Watson Health ML program uses predictive matching algorithms to dynamically combine identical and vastly different individuals, procedures, and results instead of physician experts searching the cumbersome ES-system rule data. If another medical file of the clinic is correct and truthful, preferably the Watson health care system will conclude the basic principles mentioned above (e.g., "unless the client then weighs more than or is adequate to 100 kg ..."), the ML system is in fact completely connected into a rectify and useful tool in developing. For example, if Cleveland Clinic makes more pragmatic decisions about the primary weight group, if patients over 80 kg are typically placed with the 100 kg clients, so the ML system can choose the lower threshold, or even the ML system can establish a complex series of guidelines, which may loosely be described as "if the individual is on the verge of or over 300 pounds." By trying to serve a consultation or watchdog function, CDSSs can assist physicians, patients or other customers. In the advisory position, the CDSS may indicate professional standards for post-operative patient discharge, including recommended drugs, exercise schemes, and constant follow-up checks and evaluations to ensure optimum results. CDSS can help individuals choose alternative treatment or recovery decisions, and/or assist in the safest and perhaps most cost-effective way feasible with one of the most suitable escalation and diagnosis of newly emerging health problems. CDSS may also be effective in reducing medical errors, particularly patient errors. Many EHR operating systems have qualified subsystems for protecting patients against drug errors. These CDSSs will test prescription medications against established patient reactions, possible drug–drug conflicts with current treatment regimens, and/or call for consideration for safer or less harmful medication substitutes. CDSS provides many benefits to physicians, medical administrators, and professionals alike. For example, doctors and nurses may add or amend ideas to ensure they propose new guidelines and practices. CDSS could also provide nurses and health care professionals with very invaluable help by sustaining instant access to huge databases of unordinary illnesses, therapies and problems. By the start of the Ebola outbreak (2014–15), CDSS altered to inquire and study recent global travel habits for patient populations can add Ebola to the group of possible diseases and coverage if a patient abruptly obtained headache, stomach pain, and severe fever from the emergency room. A different type of CDSS in healthcare is used in some medical equipment such as prostate cancer tumor scanning, blood testing analytics applications, but a few medical implants such as embedded defibrillators, and a few stand-alone devices such as AEDs.

6.5 ADVANCEMENTS OF AI IN MEDICAL FIELD

Artificial intelligence effectively analyses knowledge, patient data, and structure, and improves automated technology to produce quicker and more reliable performance. This technology is of interest to the individual for remote consultation and careful treatment of prescription (Haleem, Vaishya, Javaid, and Khan, 2020; Caocci et al. 2010. It lets doctors obtain better outcomes that are listed underneath:

6.5.1 RADIOLOGY

AI-assisted surgery increases precision, accuracy as well as in real-time operating environments they also understand. They could help the surgeon press for better outcomes in surgery and treatment. All the changes and improvement are helping the patient get a vibrant recovery and improved surgical chances. This technology can also pre-set the data variable associated with the pre-defined notions method, after setting it. Recently, Ai makes substantial advances in perception (the processing of sensory input) that allow significant tasks to be properly represented and interpreted.

6.5.2 CARDIOLOGY

In cardiology, AI is also used to reduce the risk of premature heart failure. It incorporates data based on information related to heart disorders. This system will make the obstruction inside the heart valve aware to prevent an attack's probability. In addition, it provides right information on blood circulation. Implementing AI helps to join a patient in any way to a hospital to seek medication until it is fully safe.

6.5.3 SURGERY

Surgeons effectively integrate AI into the surgery by collecting data from all phases. Features a bright outlook for delivering health services with the very highest level. It produces a medical judgment supported by evidence to improve the workflow of the patient care and the surgeon. It effectively offers advanced surgery with good outcomes.

6.5.4 HOSPITAL ADMINISTRATION

This system preserves the record in digital form in the healthcare sector to boost performance and accuracy. AI results in greater and synchronized variables and data in clinic management solutions as opposed to enhanced medical records, customer mechanization, and clinician data storage, catalogue, and outcomes. This software allows track clinician statistics etc and offers the physician with real-time information about the clients' family. It is also important to check treatment services, which drives the hospital successfully. This forecasts the cause behind an individual's disease correctly. AI offers remote hospital management power to improve the productivity of physicians, nurses, and employees at the facility.

Artificial intelligence will check the findings of a patient's test and update/remind the physician at the appropriate time. Areas of progress include automated electrocardiogram (ECG), heart monitoring, surgical laboratory examination, diagnostic imaging, electroencephalography, respiratory screening, and general anesthesia. This new technology will examine the biopsy, amount of sugar, medical picture, and various other tasks easily.

When patient data are encoded in formulas, AI can extract the information listed to uncover given medical issue. The device can comprehend human speech and writing through using machine learning and artificial intelligence to handle and analyze the client using various technologies. It directs doctor, pathologist, and physician on how to improve results in real-time as well as gain knowledge the talent. To gain progress and better performance, AI guides the surgeon step-by-step and more study. This also reveals what kinds of innovations are being made in the medical field. Figure 6.5 gives a deep understanding of AI process.

6.5.5 OTHER RECENT ADVANCEMENTS IN HEALTHCARE

AI and ML are now rising with the growing population to provide new and improved ways of detecting illnesses, diagnosing disorders, crowdsourcing, and designing recovery strategies, tracking health pandemics, contract performance in medical testing and clinical trials, and making operations more competitive in coping with increasing competition. In healthcare, there are different methods of using computers and computational modeling for diagnostics, watching scans and patient information, operations, and much more.

The technologies offer quantitative knowledge that replaces the healthcare industry's balancing act.

6.6 CONCLUSION

Hence, after reviewing the advancements in the field of medical science for artificial intelligence, the following can be deduced:

- Artificial intelligence can help doctors analyze patient reports and deduce an outcome that is both accurate and fast.

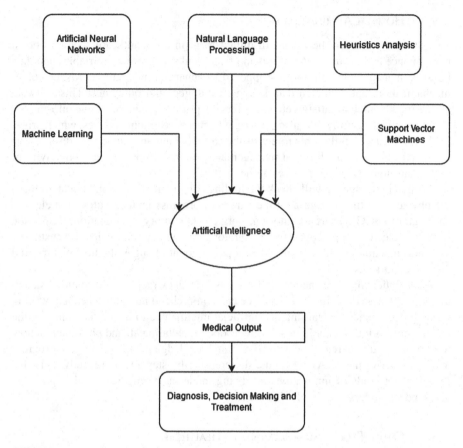

FIGURE 6.5   Process chart for AI.

- Improvement in the areas of natural language processing can lead to a future where a remote hospital or clinic can be completely operated with artificial intelligence.
- Major adoption of CDSS (Clinical Decision Support System) could help newly trained doctors by taking the load off them by advising and helping them in serious medical cases.
- Adoption of artificial intelligence by large hospitals can help them leverage the power of AI in both storing and classifying patient data for future use.

REFERENCES

Aoyama, M., Li, Q., Katsuragawa, S., MacMahon, H., and Doi, K. 2002. Automated Computerized Scheme for Distinction between Benign and Malignant Solitary Pulmonary Nodules on Chest Images. *Medical Physics*, *29*(5): 701–8. doi: 10.1118/1.1469630.

Arnold, M. 1996. Non-invasive Glucose Monitoring. *Current Opinion in Biotechnology*, *7*(1): 46–9. doi: 10.1016/s0958-1669(96)80093-0.

Asada, N., Doi, K., MacMahon, H., Montner, S., Giger, M., Abe, C., and Wu, Y. 1990. Potential Usefulness of an Artificial Neural Network for Differential Diagnosis of Interstitial Lung Diseases: Pilot Study. *Radiology*, *177*(3): 857–60. doi: 10.1148/radiology.177.3.2244001.

Ashizawa, K., Ishida, T., MacMahon, H., Vyborny, C., Katsuragawa, S., and Doi, K. 1999. Artificial Neural Networks in Chest Radiography: Application to the Differential Diagnosis of Interstitial Lung Disease. *Academic Radiology*, *6*(1): 2–9. doi: 10.1016/s1076-6332(99)80055-5.

Ashizawa, K., MacMahon, H., Ishida, T., Nakamura, K., Vyborny, C., Katsuragawa, S., and Doi, K. 1999. Effect of an Artificial Neural Network on Radiologists' Performance in the Differential Diagnosis of Interstitial Lung Disease Using Chest Radiographs. *American Journal of Roentgenology*, *172*(5), 1311–15. doi: 10.2214/ajr.172.5.10227508.

Atkov, O., Gorokhova, S., Sboev, A., Generozov, E., Muraseyeva, E., Moroshkina, S., and Cherniy, N. 2012. Coronary Heart Disease Diagnosis by Artificial Neural Networks Including Genetic Polymorphisms and Clinical Parameters. *Journal of Cardiology*, *59*(2): 190–4. doi: 10.1016/j.jjcc.2011.11.005.

Baker, J., Kornguth, P., Lo, J., Williford, M., and Floyd, C. 1995. Breast Cancer: Prediction with Artificial Neural Network Based on BI-RADS Standardized Lexicon. *Radiology*, *196*(3): 817–22. doi: 10.1148/radiology.196.3.7644649.

Birdwell, R., Bandodkar, P., and Ikeda, D. 2005. Computer-Aided Detection with Screening Mammography in a University Hospital Setting. *Radiology*, *236*(2): 451–7. doi: 10.1148/radiol.2362040864.

Caocci, G., Baccoli, R., Vacca, A., Mastronuzzi, A., Bertaina, A., and Piras, E. et al. 2010. Comparison between an Artificial Neural Network and Logistic Regression in Predicting Acute Graft-vs-Host Disease after Unrelated Donor Hematopoietic Stem Cell Transplantation in Thalassemia Patients, *Experimental Hematology*, *38*(5): 426–33. doi: 10.1016/j.exphem.2010.02.012.

Chan, K., Ling, S., Dillon, T., and Nguyen, H. 2011. Diagnosis of Hypoglycemic Episodes Using a Neural Network-Based Rule Discovery System. *Expert Systems with Applications*, *38*(8): 9799–9808. doi: 10.1016/j.eswa.2011.02.020.

Chung, H. M., Johnson, P., and Todd, P. 1997. Introduction: Expertise and Modeling Expert Decision Making. *Decision Support Systems*, *21*(2): 49–50. Doi: 10.1016/s0167-9236(97)00016-x.

Cupples, T., Cunningham, J., and Reynolds, J. 2005. Impact of Computer-Aided Detection in a Regional Screening Mammography Program. *American Journal of Roentgenology*, *185*(4): 944–50. doi: 10.2214/ajr.04.1300.

Dazzi, D., Taddei, F., Gavarini, A., Uggeri, E., Negro, R., and Pezzarossa, A. 2001. The Control of Blood Glucose in the Critical Diabetic Patient. *Journal of Diabetes and Its Complications*, *15*(2): 80–7. doi: 10.1016/s1056-8727(00)00137-9.

Dean, J., and Ilvento, C. 2006. Improved Cancer Detection Using Computer-Aided Detection with Diagnostic and Screening Mammography: Prospective Study of 104 Cancers. *American Journal of Roentgenology*, *187*(1): 20–8. doi: 10.2214/ajr.05.0111.

Destounis, S., DiNitto, P., Logan-Young, W., Bonaccio, E., Zuley, M., and Willison, K. 2004. Can Computer-aided Detection with Double Reading of Screening Mammograms Help Decrease the False-Negative Rate? Initial Experience. *Radiology*, *232*(2): 578–84. doi: 10.1148/radiol.2322030034.

Doi, K. 2003. Computer Aided Diagnosis in Chest Radiography. *Academic Radiology*, *10*(8): 944. doi: 10.1016/s1076-6332(03)00102-8.

Doi, K. 2004. Overview on Research and Development of Computer-Aided Diagnostic Schemes. *Seminars in Ultrasound, CT And MRI*, *25*(5): 404–10. doi: 10.1053/j.sult.2004.02.006.

Doi, K. 2005. Current Status and Future Potential of Computer-Aided Diagnosis in Medical Imaging. *The British Journal of Radiology*, *78*(suppl_1): s3–s19. doi: 10.1259/bjr/82933343.

Doi, K. 2006. Diagnostic Imaging over the Last 50 Years: Research and Development in Medical Imaging Science and Technology. *Physics in Medicine and Biology*, *51*(13): R5–R27. doi: 10.1088/0031-9155/51/13/r02.

Doi, K. 2007. Computer-Aided Diagnosis in Medical Imaging: Historical Review, Current Status and Future Potential. *Computerized Medical Imaging and Graphics*, *31*(4–5): 198–211. doi: 10.1016/j.compmedimag.2007.02.002.

Duda, R., and Shortliffe, E. 1983. Expert Systems Research. *Science*, *220*(4594): 261–8. doi: 10.1126/science.6340198.

Elstein, A. 2000. Clinical Problem Solving and Decision Psychology. *Academic Medicine*, *75*(Supplement): S134–6. Doi: 10.1097/00001888-200010001-00042.

Feinstein, A. 1973. Computer Diagnosis and Diagnostic Methods. *JAMA: The Journal of the American Medical Association*, *225*(1): 68. Doi:10.1001/jama.1973.03220280056034.

Freer, T., and Ulissey, M. 2001. Screening Mammography with Computer-aided Detection: Prospective Study of 12,860 Patients in a Community Breast Center. *Radiology*, *220*(3): 781–6. doi: 10.1148/radiol.2203001282.

Gilbert, F., Astley, S., McGee, M., Gillan, M., Boggis, C., Griffiths, P., and Duffy, S. 2006. Single Reading with Computer-aided Detection and Double Reading of Screening Mammograms in the United Kingdom National Breast Screening Program. *Radiology*, *241*(1): 47–53. doi: 10.1148/radiol.2411051092.

Gobry, G. 1973. Computer-Assisted Clinical Decision-Making. *Methods of Information in Medicine*, *12*(1): 45–51. doi: 10.1055/s-0038-1636093.

Gur, D., Sumkin, J., Rockette, H., Ganott, M., Hakim, C., and Hardesty, L. et al. 2004. Changes in Breast Cancer Detection and Mammography Recall Rates After the Introduction of a Computer-Aided Detection System. *JNCI Journal of the National Cancer Institute*, *96*(3): 185–90. doi: 10.1093/jnci/djh067.

Haleem, A., Vaishya, R., Javaid, M., and Khan, I. 2020. Artificial Intelligence (AI) Applications in Orthopaedics: An Innovative Technology to Embrace. *Journal of Clinical Orthopaedics and Trauma*, *11*: S80–1. doi: 10.1016/j.jcot.2019.06.012.

Henning, K. 2020. Retrieved July 13, 2020, from www.jstor.org/stable/23315680

Kano, A., Doi, K., MacMahon, H., Hassell, D., and Giger, M. 1994. Digital Image Subtraction of Temporally Sequential Chest Images for Detection of Interval Change. *Medical Physics*, *21*(3): 453–61. doi: 10.1118/1.597308.

Karabulut, E., and İbrikçi, T. 2011. Effective Diagnosis of Coronary Artery Disease Using the Rotation Forest Ensemble Method. *Journal of Medical Systems*, *36*(5): 3011–18. doi: 10.1007/s10916-011-9778-y.

Kassirer, J. 1978. Clinical Problem Solving: A Behavioral Analysis. *Annals of Internal Medicine*, *89*(2): 245. Doi: 10.7326/0003-4819-89-2-245.

Kissick, W. 1969. *Introduction to Medical Decision making*. Medical Care, *7*(6): 488–9. doi: 10.1097/00005650-196911000-00009.

Kuipers, B., and Kassirer, J. 1984. Causal Reasoning in Medicine: Analysis of a Protocol. *Cognitive Science*, *8*(4): 363–85. Doi: 10.1207/s15516709cog0804_3.

Leon, B., Alanis, A., Sanchez, E., Ornelas-Tellez, F., and Ruiz-Velazquez, E. 2012. Inverse Optimal Neural Control of Blood Glucose Level for Type 1 Diabetes Mellitus Patients. *Journal of The Franklin Institute*, *349*(5): 1851–70. doi: 10.1016/j.jfranklin.2012.02.011.

Li, Q., Katsuragawa, S., and Doi, K. 2000. Improved Contralateral Subtraction Images by Use of Elastic Matching Technique. *Medical Physics*, *27*(8): 1934–42. doi: 10.1118/1.1287112.

Lo, J., Baker, J., Kornguth, P., Iglehart, J., and Floyd, C. 1997. Predicting Breast Cancer Invasion with Artificial Neural Networks on the Basis of Mammographic Features. *Radiology*, *203*(1): 159–63. doi: 10.1148/radiology.203.1.9122385.

Lodwick, G., Haun, C., Smith, W., Keller, R., and Robertson, E. 1963. Computer Diagnosis of Primary Bone Tumors. *Radiology*, *80*(2): 273–5. doi: 10.1148/80.2.273.

Malin, J. 2013. Envisioning Watson as a Rapid-Learning System for Oncology. *Journal of Oncology Practice*, *9*(3): 155–7. doi: 10.1200/jop.2013.001021.

Matsuki, Y., Nakamura, K., Watanabe, H., Aoki, T., Nakata, H., Katsuragawa, S., and Doi, K. 2002. Usefulness of an Artificial Neural Network for Differentiating Benign from Malignant Pulmonary Nodules on High-Resolution CT. *American Journal of Roentgenology*, *178*(3): 657–63. doi: 10.2214/ajr.178.3.1780657.

Miller, R., Pople, H., and Myers, J. 1982. Internist-I, an Experimental Computer-Based Diagnostic Consultant for General Internal Medicine. *New England Journal of Medicine*, *307*(8): 468–76. doi: 10.1056/nejm198208193070803.

Morton, M., Whaley, D., Brandt, K., and Amrami, K. 2006. Screening Mammograms: Interpretation with Computer-aided Detection—Prospective Evaluation. *Radiology*, *239*(2): 375–83. doi: 10.1148/radiol.2392042121.

Nakamura, K., Yoshida, H., Engelmann, R., MacMahon, H., Katsuragawa, S., and Ishida, T. et al. 2000. Computerized Analysis of the Likelihood of Malignancy in Solitary Pulmonary Nodules with Use of Artificial Neural Networks. *Radiology*, *214*(3): 823–30. doi: 10.1148/radiology.214.3.r00mr22823.

Narasinga Rao, M., Sridhar, G., Madhu, K., and Rao, A. 2010. A Clinical Decision Support System Using Multi-Layer Perceptron Neural Network to Predict Quality of Life in Diabetes. *Diabetes and Metabolic Syndrome: Clinical Research and Reviews*, *4*(1): 57–9. doi: 10.1016/j.dsx.2009.04.002.

Özbay, Y. 2008. A New Approach to Detection of ECG Arrhythmias: Complex Discrete Wavelet Transform Based Complex Valued Artificial Neural Network. *Journal of Medical Systems*, *33*(6): 435–45. doi: 10.1007/s10916-008-9205-1.

Pauker, S., Kassirer, J., Schwartz, W., and Gorry, G. 1976. Towards The Simulation of Clinical Cognition. Taking a Present Illness by Computer. *The American Journal of Medicine*, *60*(7): A77. Doi:10.1016/0002-343(76)90591-x.

Sahiner, B., Heang-Ping Chan, Petrick, N., Datong Wei, Helvie, M., Adler, D., and Goodsitt, M. 1996. Classification of Mass and Normal Breast Tissue: A Convolution Neural Network Classifier with Spatial Domain and Texture Images. *IEEE Transactions on Medical Imaging*, *15*(5): 598–610. doi: 10.1109/42.538937.

Schwartz, W. 1970. Medicine and the Computer. *New England Journal of Medicine*, 283(23): 1257–64. Doi: 10.1056/n ejm197012032832305.

Shortliffe, E. 1976. Computer-Based Medical Consultations: MYCIN. *Annals of Internal Medicine*, *85*(6): 831. Doi: 10.7326/0003-4819-85-6-831_1.

Trajanoski, Z., Regittnig, W., and Wach, P. 1998. Simulation Studies on Neural Predictive Control of Glucose Using the Subcutaneous Route. *Computer Methods and Programs in Biomedicine*, *56*(2): 133–9. doi: 10.1016/s0169-2607(98)00020-0.

Uğuz, H. 2010. A Biomedical System Based on Artificial Neural Network and Principal Component Analysis for Diagnosis of the Heart Valve Diseases. *Journal of Medical Systems*, *36*(1): 61–72. doi: 10.1007/s10916-010-9446-7.

Weiss, S., Kulikowski, C., Amarel, S., and Safir, A. 1978. A Model-Based Method for Computer-Aided Medical Decision-Making. *Artificial Intelligence*, *11*(1–2): 145–72. doi: 10.1016/0004-3702(78)90015-2.

Wu, Y., Doi, K., Giger, M., Metz, C., and Zhang, W. 1994. Reduction of False Positives in Computerized Detection of Lung Nodules in Chest Radiographs Using Artificial

Neural Networks, Discriminant Analysis, and a Rule-Based Scheme. *Journal of Digital Imaging*, 7(4): 196–207. doi: 10.1007/bf03168540.

Wu, Y., Giger, M., Doi, K., Vyborny, C., Schmidt, R., and Metz, C. 1993. Artificial Neural Networks in Mammography: Application to Decision Making in the Diagnosis of Breast Cancer. *Radiology*, *187*(1): 81–7. doi: 10.1148/radiology.187.1.8451441.

Zhang, W., Doi, K., Giger, M., Wu, Y., Nishikawa, R., and Schmidt, R. 1994. Computerized Detection of Clustered Microcalcifications in Digital Mammograms Using a Shift-Invariant Artificial Neural Network. *Medical Physics*, *21*(4): 517–24. doi: 10.1118/1.597177.

7 Machine Learning Approach for Breast Cancer Early Diagnosis

*Sangeeta Mangesh, Vishesh Saxena,
Utkarsh Tripathi, and Yamini Gupta*

CONTENTS

7.1 INTRODUCTION

Breast cancer is a global disease. Many European countries and the USA have in fact recommended that every female citizen must undergo mammogram testing biannually especially for the age group of 50–74. As per the WHO web page "Breast cancer is top cancer in women both in the developed and the developing world" (Ogbu and Arah 2016) the gravity of health concerns is more in developing countries due to lifestyle changes, urbanization, pollution, and lack of medical infrastructure. To mitigate health concerns, the only way out is adaptation of technology for reaching out to masses and creating awareness for early detection of symptomatic patients. In breast cancer patients the survival rate greatly depends upon the stage of advancement or cancer metastatic condition. Early detection and diagnosis is a windfall in improving quality of life as well as increasing life expectancy of these patients (Ogbu and Arah 2016; Akundi et al. 2020; BCI 2016; Fitzmaurice et al. 2015.

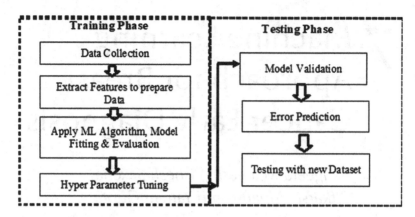

FIGURE 7.1   Block schematic representation of the research flow.

This research is aimed at developing a technology aided breast cancer early diagnostic application that will help the mankind. As there is a huge volume of data of the different features extracted from the mammographic images is available; it can significantly contribute towards a decision support system (DSS) for an early detection of breast cancer detection. Through this chapter a possibility of such system is explored using machine learning (ML) and artificial intelligence (AI).

With the advancement in the data science possibility of classification of the feature using machine learning algorithms has been explored by the researchers (Gardezi et al. 2019; Eleyan 2012; Asri et al. 2016; Atrey et al. 2019). The majority of the researchers have used the Wisconsin Diagnostic Breast Cancer (WDBC) dataset (Vig 2014; Agarap 2018; Atrey et al. 2019; Asri et al. 2016; Omondiagbe, Veeramani, and Sidhu 2019; Nallamala, Mishra, and Koneru 2019; Eleyan 2012) and eventually presented significant results.

The highlighting feature of this research is a step forward with ten different machine learning classification results. The results of the machine learning classifiers namely Logistic Regression (LR), K-Nearest Neighbors(KNN), Decision Tree (DT), Random Forest(RF), Quadratic Discriminant Analysis (QDA), Naïve Bayes (NV), Support Vector Machine (SVM), Radial Basis Function (RBF), AdaBoost Tree (ABT), and Multilayer Perceptron (MLP) (Kotsiantis, Zaharakis, and Pintelas 2006; Aggarwal 2014) ("Scikit-Learn 0.23.2," n.d.) along with optimization of the classifier using hyperparameters tuning and experimenting for certain values for achieving better accuracy.The successful model design has also been verified by performing testing of the new dataset to generate false-positive cases to estimate errors in the cancer prediction. The entire set of results thus validates a feasible DSS design that can aid in early diagnosis of this disease. The block schematic of the entire process followed through this research is represented in Figure 7.1

7.2   ABOUT THE DATASET

The data analyzed for this research is from The UCI Machine Learning Repository for the breast cancer dataset. It's a public dataset created by Dr. William H. Wolberg, a physician at the University of Wisconsin Hospital at Madison, Wisconsin, USA. Dr. Wolberg has used a graphical computer program 'Xcyte' to extract features from digital scanning of the fluid samples from patients with solid breast masses through (Fine Needle Aspiration Procedure) FNA to create this dataset (Lichman 2013; Aggarwal 2014; Akundi, Rahman, Wen, Xu, and Tseng 2020). A curve-fitting algorithm has also been used to compute ten different features from each one of the cells in the sample, then it calculates the mean value, extreme value, and standard error of each feature for the image, returning 30 real-valued vectors. WDBC database has data representation for 569 patients with 212 malignant and 357 benign cases identified (Lichman 2013). The dataset contains 32 columns with the first two attributes corresponding to the identifier number and the diagnosis status. The remaining values are the 30 real attributes, including, the mean, the standard error, and the worst of 10 cell nucleus features that include the radius, texture, perimeter, area, smoothness, compactness, concave points, concavity, symmetry, and fractal dimension. Also, the mean, standard error, and "worst" or largest (mean of the three largest values) of these features have also been estimated for each image (with no missing value), among the 30 features available in the dataset (Lichman 2013; Gutierrez 2014). For example, field 3 is Mean Radius, field 13 is Radius SE, field 23 is Worst Radius.(Lichman 2013; Tolson 2001; Nallamala, Mishra, and Koneru 2019; Gutierrez 2014).

7.3 METHODOLOGY

7.3.1 EXPERIMENTAL ENVIRONMENT

For the machine learning classification analysis or processing of the data using Python, various predefined Python libraries have been used to perform specific tasks. These include -NumPy, Pandas, Matplotlib, Scikit Learn, OS, Collections. The *Anaconda* platform is used with the *Jupyter Notebook* application to implement Python code ("The Jupyter Notebook-Documentation," n.d.). Before estimating the numerical values of various instances from the dataset, it is divided into training and testing in a ratio of 80:20. Data visualization is a key aspect of machine learning library 'pandas' visualization which is built on top of matplotlib, has been used to find the data distribution of the features (Gutierrez 2014; Lichman 2013; "The Jupyter Notebook-Documentation" n.d.; Aggarwal 2014). Figure 7.2 shows the ML and AI methodology used.

7.3.2 DATA VISUALIZATION

To generate the heat mapping features of data visualization and generate a correlation matrix, the entire dataset is categorized into three categories. First one is the "mean." The mean category contains all the mean values of all the parameters considered as features that define specific characteristic of the tumor such as mean of radius (i.e. the mean of distances from the center to points on the perimeter), texture (standard deviation of gray-scale values), perimeter (mean size of the core tumor), area, smoothness (mean of local variation in radius lengths), compactness (mean of perimeter^2 / area – 1.0), concavity (mean of the severity of concave portions of the contour), concave points (mean for a number of concave portions of the contour), symmetry and fractal

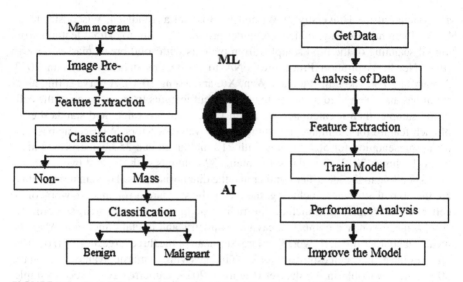

FIGURE 7.2   Both ML and AI being the highlighting feature of this research.

dimensions (mean for "coastline approximation"– 1) (Lichman 2013; Aggarwal 2014; Kotsiantis, Zaharakis, and Pintelas 2006; Gutierrez 2014; Akundi, Rahman, Wen, Xu, and Tseng 2020; Brownlee 2014; Lugat 2018. The heatmap of the mean values is shown in Figure 7.3.

The second category includes standard error values for all the parameters considered for estimating mean values and the third category includes worst case values of the same features. The heat map plots generated for these two categories for all the considered features are indicted in Figures 7.4 and 7.5 respectively (Gutierrez 2014; Aggarwal 2014; Brownlee 2014; Ferroni et al. 2019).

After performing the of the heat map plots to extracting the features that can lead to prognosis of breast cancer following observations have been noted:

- The heat map plot of mean values: The mean values of radius, perimeter, and area are highly correlated with one another so only parameters out of them are selected. Out of compactness, concavity, and concave point the only feature, i.e. compactness, is considered as all three are found to be highly correlated. Thus, the mean values of perimeter, texture, compactness, symmetry and smoothness remain the final selected features from the first category.
- The heat map plot of the second category "standard error": As there exists heavy correlation between radius, perimeter, and area, so only one is selected out of these three features. Similarly out of compactness, concavity, and concave point only compactness is selected as these three are also heavily correlated. Thus, the finally selected features from the second category i.e. standard error parameters, are perimeter, texture, compactness, symmetry, and smoothness.
- The third category "worst" parameter: The radius, perimeter, and area are again heavily correlated; leading to selection of only one feature out of them.

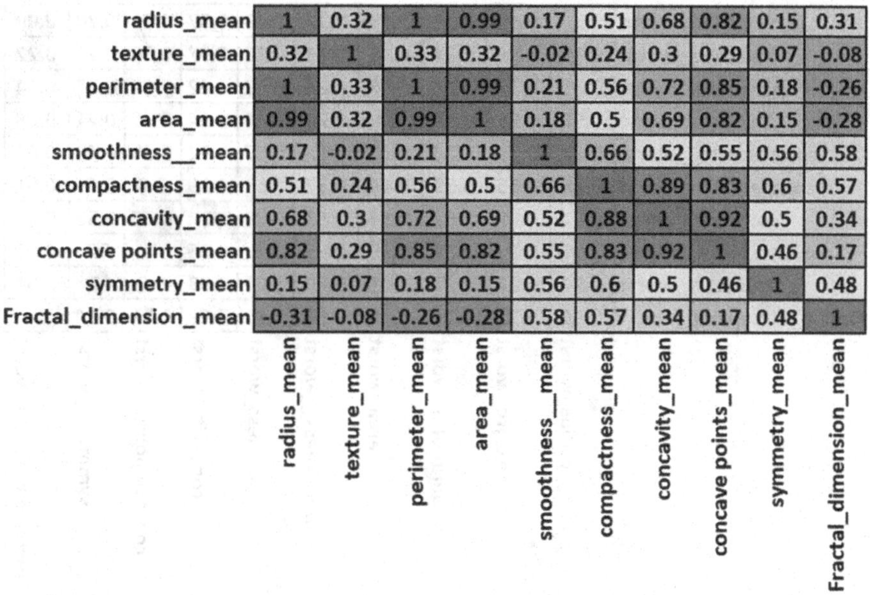

	radius_mean	texture_mean	perimeter_mean	area_mean	smoothness_mean	compactness_mean	concavity_mean	concave points_mean	symmetry_mean	Fractal_dimension_mean
radius_mean	1	0.32	1	0.99	0.17	0.51	0.68	0.82	0.15	0.31
texture_mean	0.32	1	0.33	0.32	-0.02	0.24	0.3	0.29	0.07	-0.08
perimeter_mean	1	0.33	1	0.99	0.21	0.56	0.72	0.85	0.18	-0.26
area_mean	0.99	0.32	0.99	1	0.18	0.5	0.69	0.82	0.15	-0.28
smoothness__mean	0.17	-0.02	0.21	0.18	1	0.66	0.52	0.55	0.56	0.58
compactness_mean	0.51	0.24	0.56	0.5	0.66	1	0.89	0.83	0.6	0.57
concavity_mean	0.68	0.3	0.72	0.69	0.52	0.88	1	0.92	0.5	0.34
concave points_mean	0.82	0.29	0.85	0.82	0.55	0.83	0.92	1	0.46	0.17
symmetry_mean	0.15	0.07	0.18	0.15	0.56	0.6	0.5	0.46	1	0.48
Fractal_dimension_mean	-0.31	-0.08	-0.26	-0.28	0.58	0.57	0.34	0.17	0.48	1

FIGURE 7.3 Heatmap of the mean values.

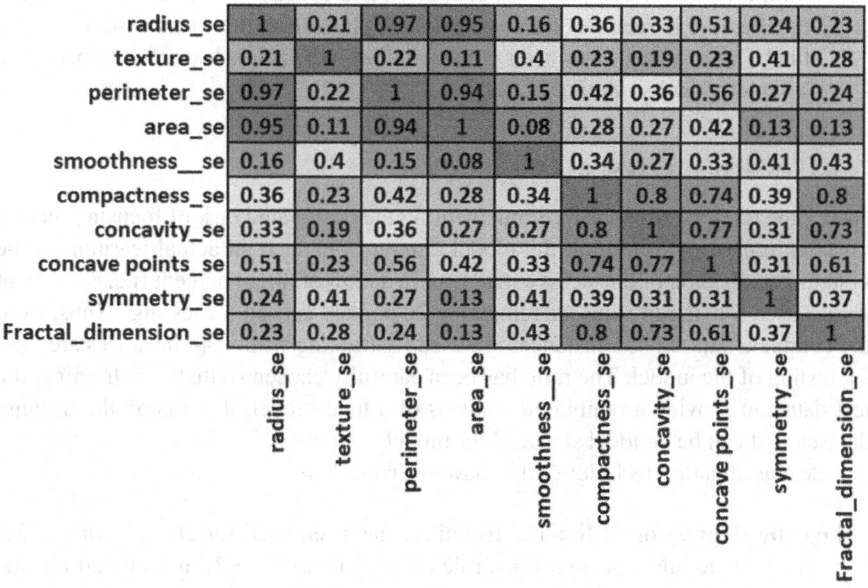

	radius_se	texture_se	perimeter_se	area_se	smoothness_se	compactness_se	concavity_se	concave points_se	symmetry_se	Fractal_dimension_se
radius_se	1	0.21	0.97	0.95	0.16	0.36	0.33	0.51	0.24	0.23
texture_se	0.21	1	0.22	0.11	0.4	0.23	0.19	0.23	0.41	0.28
perimeter_se	0.97	0.22	1	0.94	0.15	0.42	0.36	0.56	0.27	0.24
area_se	0.95	0.11	0.94	1	0.08	0.28	0.27	0.42	0.13	0.13
smoothness__se	0.16	0.4	0.15	0.08	1	0.34	0.27	0.33	0.41	0.43
compactness_se	0.36	0.23	0.42	0.28	0.34	1	0.8	0.74	0.39	0.8
concavity_se	0.33	0.19	0.36	0.27	0.27	0.8	1	0.77	0.31	0.73
concave points_se	0.51	0.23	0.56	0.42	0.33	0.74	0.77	1	0.31	0.61
symmetry_se	0.24	0.41	0.27	0.13	0.41	0.39	0.31	0.31	1	0.37
Fractal_dimension_se	0.23	0.28	0.24	0.13	0.43	0.8	0.73	0.61	0.37	1

FIGURE 7.4 Heatmap for parameter SE.

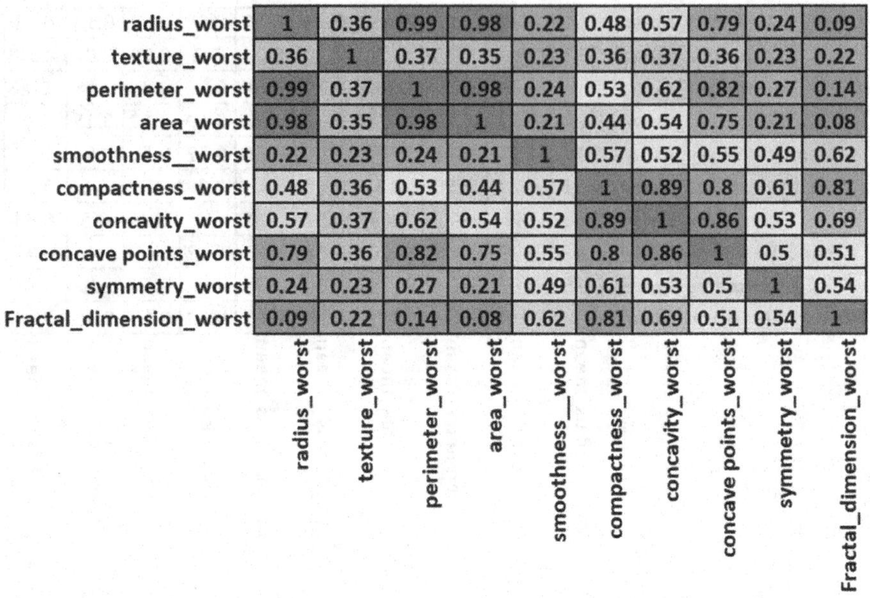

	radius_worst	texture_worst	perimeter_worst	area_worst	smoothness_worst	compactness_worst	concavity_worst	concave points_worst	symmetry_worst	Fractal_dimension_worst
radius_worst	1	0.36	0.99	0.98	0.22	0.48	0.57	0.79	0.24	0.09
texture_worst	0.36	1	0.37	0.35	0.23	0.36	0.37	0.36	0.23	0.22
perimeter_worst	0.99	0.37	1	0.98	0.24	0.53	0.62	0.82	0.27	0.14
area_worst	0.98	0.35	0.98	1	0.21	0.44	0.54	0.75	0.21	0.08
smoothness__worst	0.22	0.23	0.24	0.21	1	0.57	0.52	0.55	0.49	0.62
compactness_worst	0.48	0.36	0.53	0.44	0.57	1	0.89	0.8	0.61	0.81
concavity_worst	0.57	0.37	0.62	0.54	0.52	0.89	1	0.86	0.53	0.69
concave points_worst	0.79	0.36	0.82	0.75	0.55	0.8	0.86	1	0.5	0.51
symmetry_worst	0.24	0.23	0.27	0.21	0.49	0.61	0.53	0.5	1	0.54
Fractal_dimension_worst	0.09	0.22	0.14	0.08	0.62	0.81	0.69	0.51	0.54	1

FIGURE 7.5   Heatmap for parameter worst.

Similarly, out of compactness, concavity, and concave point only the compactness feature is selected due to correlation. Thus, from the worst case parameter final features that have been considered are perimeter, texture, compactness, symmetry, and smoothness.

7.3.3   MACHINE LEARNING CLASSIFICATION

As per the strategy adopted for implementing the designated task of focusing mainly towards imparting training to the model aiming to happen an actual learning of the system, the division of dataset is carried out in a ratio of 80:20 percent (i.e. 80 percent data is used for training and the remaining 20 percent is used for testing). Thus, from the complete dataset, 455 instances are used for training while 144 instances are used for testing of the model. The ratio has been carefully chosen so that post-training, the test data can provide an unbiased analysis of a final model, that fits on the training dataset, and can be an ideal standard for model evaluation.

The classification techniques that have been used are:

Logistic Regression (LR): The technique has been used for classification tasks. The logistic rule is a supervised rule ("Scikit-Learn 0.23.2," n.d.) that trains the model by taking input variables and a target variable. This algorithm has been applied to train and test our model and find the accuracy and cross-validation score by using the existing functions available in the package (Galvan et al. 2016;

Omondiagbe, Veeramani, and Sidhu 2019; Hussain et al. 2018; Brownlee 2014; Lichman 2013; Vig 2014; Asri et al. 2016; Laghmati, Tmiri, and Cherradi 2019).

K-Nearest Neighbors (KNN): This algorithm is used with default neighbor value of 15 and power parameter for the Minkowski metric to be 2 ("Scikit-Learn 0.23.2," n.d.). KD Tree algorithm has been used with a leaf size of 20 to evaluate accuracy and cross-validation score (Nallamala, Mishra, and Koneru 2019; Gardezi et al. 2019; Eleyan 2012; Omondiagbe, Veeramani, and Sidhu 2019; Ming et al. 2019; Laghmati, Tmiri, and Cherradi 2019).

Decision Tree (DT): This is the most prevalent method of supervised learning used for classification. The best strategy of node splitting has also been incorporated with criteria gini (Nallamala, Mishra, and Koneru 2019; Shen et al. 2019; Gardezi et al. 2019; Tolson 2001; Vig 2014; Ming et al. 2019).

A random forest (RF): This is a meta estimator that fits a number of decision tree classifiers on various sub-samples of the dataset and uses averaging to improve the predictive accuracy and control over-fitting (Aggarwal 2014; "Scikit-Learn 0.23.2," n.d.). The number of trees has been optimized to 100. This has been used by importing parameters using the sklearn ensemble ("Scikit-Learn 0.23.2," n.d.). Random Forest Classifier method (Nallamala, Mishra, and Koneru 2019; Shen et al. 2019).

Naive Bayes (NV): This method involves a group of classification algorithms supported Bayes' Theorem. For this research, data classification has been done using functionalities available such as sklearn.naive_bayes. GaussianNB method. ("Scikit-Learn 0.23.2," n.d.) (Aggarwal 2014). The parameter optimized during implementation includes: var_smoothing=1e-7. The portion of the largest variance of all features is added to variances for calculation stability (Gardezi et al. 2019; Agarap 2018).

An AdaBoost classifier (ABT): (Aggarwal 2014; Kotsiantis, Zaharakis, and Pintelas 2006; "Scikit-Learn 0.23.2," n.d.). It's a classifier having meta estimation capability that begins with fitting of the classifier on the original dataset and then by fitting additional copies of the classifier on the same dataset but where the weights of incorrectly classified instances are adjusted such that subsequent classifiers focus more on difficult cases. The optimization parameters values of learning rate as 1 and number of estimators to terminate boosting has been set to 20 for the analysis (Dhahri et al. 2019; Asri et al. 2016; Ferroni et al. 2019).

Radial basis function (RBF): ("Scikit-Learn 0.23.2," n.d.; Gutierrez 2014; "The Jupyter Notebook-Documentation," n.d.; Aggarwal 2014). One of the important algorithms used in learning and has been implemented using the built in functionality: sklearn.gaussian_process.kernels. RBF method (Omondiagbe, Veeramani, and Sidhu 2019; Vig 2014) ("Scikit-Learn 0.23.2," n.d.).

Support Vector Machine (SVM): ("Scikit-Learn 0.23.2," n.d.) This is a supervised machine learning algorithmic rule used for classification with regularization parameter values as 1 and kernel to be linear (Kourou et al. 2015; Asri et al. 2016; Hussain et al. 2018; Laghmati, Tmiri, and Cherradi 2019; Agarap 2018; Vig 2014).

Quadratic Discriminant Analysis (QDA): This is a variant of LDA in which an individual covariance matrix is estimated for every class of observations. The estimation of regularizing the per-class covariance is set to 1 (Atrey et al. 2019; Omondiagbe, Veeramani, and Sidhu 2019; Vig 2014; Ming et al. 2019).

Multi-Layer Perceptron (MLP): It is a class of feedforward artificial neural networks (ANN) ("Scikit-Learn 0.23.2," n.d.). The specifications used for implementing the classification include: learning rate as 1, number of hidden nodes to be 32, maximum iterations to be 1000, adam optimization-based weight optimization, with hidden layer activation function to be regular linear (Agarap 2018).

7.3.4   PARAMETERS USED TO GENERATE CLASSIFICATION REPORT

The Classification Report in machine learning has been generated using the sklearn. metrics.classification_report method ("Scikit-Learn 0.23.2," n.d.). The parameters include:

Accuracy (Aggarwal 2014; Khourdifi and Bahaj 2019; Ferroni et al. 2019): Accuracy is the fraction of prediction (Kotsiantis, Zaharakis, and Pintelas 2006; Aggarwal 2014). It determines the number of correct predictions over the total number of predictions made by the model and estimated as (Ferroni et al. 2019; Kourou et al. 2015; Vig 2014; Omondiagbe, Veeramani, and Sidhu 2019; Brownlee 2014):

$$Accuracy = \frac{TP + TN}{TP + TN + FP + FN} \tag{7.1}$$

Recall: It is a measure of the proportion of patients that were predicted to have complications among those patients that have complications (Kotsiantis, Zaharakis, and Pintelas 2006; Aggarwal 2014; Akundi, Rahman, Wen, Xu, and Tseng 2020). Recall can be calculated as (Brownlee 2014):

$$Recall = \frac{TP}{TP + TN} \tag{7.2}$$

Precision: It is described as a measure of the proportion of patients that have complications among those classified to have complications by the model (Kotsiantis, Zaharakis, and Pintelas 2006; Aggarwal 2014; Akundi, Rahman, Wen, Xu, and Tseng 2020). The formula for Precision is (Brownlee 2014):

$$Precision : \frac{TP}{TP + FP} \tag{7.3}$$

F1 Score (Aggarwal 2014)**:** Weighted average of precision and recall is known as the F1 score. Considering, false positives and false negatives it is estimated. Intuitively it is not as simple to grasp as accuracy, however, F1 typically comes handy during classification in addition to accuracy (Brownlee 2014):

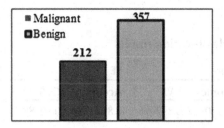

FIGURE 7.6 Diagnostic data of benign Vs malignancy from Wisconsin dataset.

$$F1 \, Score = \frac{Precision * Recall}{Precision + Recall} * 2 \qquad (7.4)$$

Cross-validation score (Aggarwal 2014): Cross-validation is a statistical method used to estimate the skill of machine learning models. The k-fold cross-validation is a procedure used to estimate the skill of the model on new data (Brownlee 2014).

7.3.5 RESULTS OF THE WISCONSIN DATASET

The 569 instances from the dataset, 357 have been diagnosed as benign and 212 have been diagnosed as malignant. The graphical representation of the diagnosis is shown in Figure 7.6 (Vig 2014).

7.4 DISCUSSION ON RESULTS

7.4.1 MODEL PERFORMANCE

The performance has been evaluated using accuracy and cross-validation score. The resultant data after applying all these ten classifiers have been tabulated in Table 7.1.

From the classification data it has been observed that the accuracy of the Support Vector Machine Classifier (Asri et al. 2016; Kourou et al. 2015; Hussain et al. 2018; Agarap 2018) is the highest in case of worst parameter subcategory (~95.61%), so it has been identified as the classifier to be used for predicting the testing dataset model. To apply this for prediction of the cancer possibility, the model is evaluated to check the malignancy of the tumor based on the features and the outcome of the predicted data is indicated in Figure 7.7. It has been observed that the model is 95.61 percent accurate and predicting whether the cancer is M = malignant or B = benign (1 means the cancer is malignant and 0 means benign).

7.5 SUMMARY OF RESULTS

In terms of machine learning estimators considered in this research, based on outcome of an accuracy, Support Vector Machine (SVM), Logistic Regression (LR), Naïve Bayes (NB) and Quadratic Discriminant Analysis (QDA) classifiers have high scores equal to 0.9561 followed by Random Forest Tree having score of -0.9474,

TABLE 7.1
Machine learning classification results

Model	Mean		SE		Worst	
	Accuracy	CVS	Accuracy	CVS	Accuracy	CVS
Logistic Regression	0.903509	0.892496	0.894737	0.80888	0.95614	0.940705
KNN	0.868421	0.90566	0.868421	0.813324	0.938596	0.949496
Decision Tree	0.850877	0.923218	0.824561	0.760405	0.894737	0.904705
Random Forest	0.877193	0.931889	0.833333	0.802189	0.947368	0.962636
Naïve Bayes	0.894737	0.916625	0.894737	0.764993	0.95614	0.951695
AdaBoost	0.921053	0.934112	0.842105	0.82197	0.938596	0.942831
Radial Basis Function	0.850877	0.896867	0.885965	0.806683	0.929825	0.947298
Support Vector Machine	0.877193	0.912205	0.894737	0.808904	0.95614	0.958264
Quadratic Discriminant Analysis	0.912281	0.87711	0.868421	0.785876	0.95614	0.916624
Multilayer Perception	0.947368	0.730908	0.894737	0.800064	0.947368	0.834014

```
prediction

array([1, 0, 0, 0, 0, 1, 1, 0, 0, 0, 1, 0, 0, 1, 0, 0, 0, 0, 0, 0, 0, 1,
       0, 0, 1, 0, 0, 0, 0, 0, 0, 0, 1, 0, 1, 0, 0, 1, 0, 0, 0, 0, 0, 1,
       1, 0, 1, 0, 1, 0, 0, 0, 0, 0, 1, 0, 0, 1, 0, 1, 0, 1, 1, 0, 0, 0,
       1, 0, 0, 0, 0, 0, 0, 0, 0, 0, 0, 0, 1, 0, 1, 1, 0, 0, 0, 0, 1, 1,
       0, 0, 0, 0, 0, 0, 0, 0, 0, 0, 0, 0, 0, 0, 0, 0, 0, 0, 0, 1, 1, 1,
       1, 1, 1, 0])
```

FIGURE 7.7 Screenshot of the model prediction applied to the dataset

Multi-Layer Perceptron(MLP) with -0.9474 and K Nearest Neighbour(KNN) with a score of 0.9386. The Decision Tree performs the worst among all ten, having a score of 0.8947.

However, in terms of the cross-validation score SVM classifier has the highest value at 0.9582 followed by KNN at 0.9495 and Logistic Regression at 0.9407.

The cross-validation score of Random Forest Tree is having the highest score of 0.9626 but in terms of accuracy, it lags behind SVM.

The results of performance parameters are tabulated in Table 7.2 and the predicted result of testing is indicated in Table 7.3. Prediction results justify the validity of the model as the error value in detecting malignancy is minimum.

TABLE 7.2
Performance parameter score

Classification Report	Precision	Recall	F1-score	Support
0	1.00	0.94	0.97	88
1	0.84	1.00	0.91	26
Accuracy			0.96	114
Average	0.92	0.97	0.94	114
Weighted avg	0.96	0.96	0.96	114

TABLE 7.3
Results of model validation

Confusion matrix	Predicted benign	Predicted malignant
True Benign	83	5
True Malignant	0	26

Therefore, accounting for exponential runtime while handing huge dataset recall value becomes a significant score which makes the SVM model as the most suited model for training. The same classifier model has been used for testing. The classifier is expected to produce results with a loss of about 4.82. The support vector machine (SVM) classifier scores 0.91 for malignant cases (1) and 0.97 for the benign cases (0) when it comes to the F1 Score. Considering the other performance matrix into account, a lot can be determined regarding the performance of the algorithms. Support Vector Machine (SVM) scores a perfect 1.000 for malignant cases and a score of 0.94 for benign cases when it comes to recall, which is critical in terms of disease prediction. It is also observed that the accuracy and cross-validation score of different classifiers may vary while implementing the model with different datasets which suggests that the machine learning algorithms are specialized.

7.6 CONCLUSION

As the presented results mainly focused on the advancement of predictive models to achieve good accuracy in predicting valid disease outcomes using supervised machine learning methods.

From the training results, it is concluded that the Support Vector Machine (SVM) classifier performs better when it comes to Breast Cancer Prediction for this dataset used.

The early diagnosis of BC can improve the prognosis and chance of survival significantly, as it can promote timely clinical treatment to patients which can only be possible by using the Machine Learning Technique combined with AI as stated in this chapter. Based on the results of this study, the prediction of breast cancer can be done with an accuracy of 95.61 percent using the best classifier identified.

This may bring new hope towards the possibility of developing a tool with the help of an intelligent system and applying data mining tools with the capability of automatically breast cancer diagnosis and proposing the best treatment.

REFERENCES

Agarap, A. F. M. 2018. On Breast Cancer Detection: An Application of Machine Learning Algorithms on the Wisconsin Diagnostic Dataset. *ACM International Conference Proceeding Series*, 1: 5–9. https://doi.org/10.1145/3184066.3184080.

Aggarwal, C. C. 2014. *Educational and Software Resources for Data Classification. Data Classification: Algorithms and Applications*. https://doi.org/10.1201/b17320.

Akundi, S., M. F. Rahman, Y. Wen, H. Xu, and T.-L. Tseng. 2020. *Advances in Telemedicine for Health Monitoring: Technologies, Design and Applications*. https://doi.org/10.1049/pbhe023e.

Asri, H., H. Mousannif, H. Al Moatassime, and T. Noel. 2016. Using Machine Learning Algorithms for Breast Cancer Risk Prediction and Diagnosis. *Procedia Computer Science*, 83 (Fams): 1064–69. https://doi.org/10.1016/j.procs.2016.04.224.

Atrey, K., Y. Sharma, N. K. Bodhey, and B. K. Singh. 2019. Breast Cancer Prediction Using Dominance-Based Feature Filtering Approach: A Comparative Investigation in Machine Learning Archetype. *Brazilian Archives of Biology and* Technology, 62: 1–15. https://doi.org/10.1590/1678-4324-2019180486.

BCI. 2016. Breast Cancer India. *Breast Cancer Research INDIA*.

Brownlee, J. 2014. https://machinelearningmastery.com/classification-accuracy-is-not-enough-more-performance-measures-you-can-use/.

Dhahri, H., E. Al Maghayreh, A. Mahmood, W. Elkilani, and M. F. Nagi. 2019. Automated Breast Cancer Diagnosis Based on Machine Learning Algorithms.*Journal of Healthcare Engineering* 2019. https://doi.org/10.1155/2019/4253641.

Eleyan, A. 2012. Breast Cancer Classification Using Moments. *2018 Electric Electronics, Computer Science, Biomedical Engineerings' Meeting (EBBT)*, 1–4. https://doi.org/10.1109/siu.2012.6204778.

García Galvan, F. R., V. Barranco, J. C. Galvan, S. Batlle, and S. Feliu Fajardo. 2016. We Are IntechOpen, the World's Leading Publisher of Open Access Books Built by Scientists, for Scientists TOP 1%. *Intech*, i: 13. https://doi.org/http://dx.doi.org/10.5772/57353.

Ferroni, P., F. M. Zanzotto, S. Riondino, N. Scarpato, F. Guadagni, and M. Roselli. 2019. Breast Cancer Prognosis Using a Machine Learning Approach. Cancers, 11(3): 1–9. https://doi.org/10.3390/cancers11030328.

Fitzmaurice, C., D. Dicker, A. Pain, H. Hamavid, M. Moradi-Lakeh, M. F. MacIntyre, C. Allen, et al. 2015. The Global Burden of Cancer 2013. *JAMA* Oncology, 1(4): 505–27. https://doi.org/10.1001/jamaoncol.2015.0735.

Gardezi, S. J. S., A. Elazab, B. Lei, and T. Wang. 2019. Breast Cancer Detection and Diagnosis Using Mammographic Data: Systematic Review. *Journal of Medical Internet* Research, 21(7): 1–22. https://doi.org/10.2196/14464.

Gutierrez, D. D. 2014. *Machine Learning and Data Science. IGARSS 2014*. www.kaggle.com/learn/intro-to-machine-learning.

Hussain, L., W. Aziz, S. Saeed, S. Rathore, and M. Rafique. 2018. Automated Breast Cancer Detection Using Machine Learning Techniques by Extracting Different Feature Extracting Strategies. *Proceedings – 17th IEEE International Conference on Trust,*

Security and Privacy in Computing and Communications and 12th IEEE International Conference on Big Data Science and Engineering, Trustcom/BigDataSE 2018, 327–31. https://doi.org/10.1109/TrustCom/BigDataSE.2018.00057.

Khourdifi, Y., and M. Bahaj. 2019. Applying Best Machine Learning Algorithms for Breast Cancer Prediction and Classification. *2018 International Conference on Electronics, Control, Optimization and Computer Science, ICECOCS 2018*, 1–5. https://doi.org/ 10.1109/ICECOCS.2018.8610632.

Kotsiantis, S. B., I. D. Zaharakis, and P. E. Pintelas. 2006. Machine Learning: A Review of Classification and Combining Techniques. *Artificial Intelligence* Review, 26(3): 159–90. https://doi.org/10.1007/s10462-007-9052-3.

Kourou, K., T. P. Exarchos, K. P. Exarchos, M. V. Karamouzis, and D. I. Fotiadis. 2015. Machine Learning Applications in Cancer Prognosis and Prediction. *Computational and Structural Biotechnology* Journal, 13: 8–17. https://doi.org/10.1016/j.csbj.2014.11.005.

Laghmati, S., A. Tmiri, and B. Cherradi. 2019. Machine Learning Based System for Prediction of Breast Cancer Severity. *Proceedings – 2019 International Conference on Wireless Networks and Mobile Communications, WINCOM 2019*, 8–12. https://doi.org/10.1109/ WINCOM47513.2019.8942575.

Lichman, M. 2013. UCI Machine Learning Repository [http://Archive.Ics.Uci.Edu/Ml]. UCI Machine Learning Repository.

Lugat, V. 2018. www.kaggle.com/vincentlugat/breast-cancer-analysis-and-prediction.

Ming, C., V. Viassolo, N. Probst-Hensch, P. O. Chappuis, I. D. Dinov, and M. C. Katapodi. 2019. Machine Learning Techniques for Personalized Breast Cancer Risk Prediction: Comparison with the BCRAT and BOADICEA Models. *Breast Cancer Research*, 21 (1): 1–11. https://doi.org/10.1186/s13058-019-1158-4.

Nallamala, S. H., P. Mishra, and S. V. Koneru. 2019. Breast Cancer Detection Using Machine Learning Way. *International Journal of Recent Technology and* Engineering, 8 (2 Special Issue 3): 1402–5. https://doi.org/10.35940/ijrte.B1260.0782S319.

Ogbu, U. C., and O. A. Arah. 2016. World Health Organization. *International Encyclopedia of Public Health*. https://doi.org/10.1016/B978-0-12-803678-5.00499-9.

Omondiagbe, D. A., S. Veeramani, and A. S. Sidhu. 2019. Machine Learning Classification Techniques for Breast Cancer Diagnosis. *IOP Conference Series: Materials Science and Engineering*, 495(1): 0–16. https://doi.org/10.1088/1757-899X/495/1/012033.

Scikit-Learn 0.23.2. https://scikit-learn.org/stable/supervised_learning.html#supervised-learning.

Shen, L., L. R. Margolies, J. H. Rothstein, E. Fluder, R. McBride, and W. Sieh. 2019. Deep Learning to Improve Breast Cancer Detection on Screening Mammography. *Scientific Reports*, 9(1): 1–12. https://doi.org/10.1038/s41598-019-48995-4.

The Jupyter Notebook-Documentation. n.d. https://doi.org/https://jupyter-notebook.readthe docs.io/en/stable/.

Tolson, E. 2001. Machine Learning in the Area of Image Analysis and Pattern Recognition. *Advanced Undergraduate Project Spring*. http://stuff.mit.edu:8001/afs/athena/course/ urop/profit/PDFS/EdwardTolson.pdf.

Vig, L. 2014. Comparative Analysis of Different Classifiers for the Wisconsin Breast Cancer Dataset. OALib, 01(06): 1–7. https://doi.org/10.4236/oalib.1100660.

8 Intelligent Surveillance System Using Machine Learning

*Anshuman Jaiswal, Lavkush Sharma,
Akshay Kumar, and Suryansh Chauhan*

CONTENTS

8.1 INTRODUCTION

Face recognition is the most basic and advanced technology that we used in today's world and in our daily life (Owayjan et al. 2013); it is also a part of biometrics that recognizes the human face based some facial characteristics which are different in every human face. We develop the idea of this project from a present face recognition search system used in China (Eagle Eye China 2017) known as "Eagle Eye." They use CCTV camera for surveillance that has face recognition system along with a database. These cameras, installed in various streets of China, when needed are used to find or detect particular people by the Chinese police or government so as to control the crime rate and arrest absconding criminals. We have followed through this idea and used it as a feature in our project to improve present video surveillance using CCTV cameras. We have used Raspberry Pi 4 which is a small single board computer developed in the United Kingdom by the Raspberry Pi Foundation as one of the main hardware components for the image processing required in this project. We have use this to make the CCTV self capable of processing the facial recognition and other works on its own rather than the old face recognition CCTV camera which is connected to a main computer for database record and image processing. In the research work of Shilpashree et al. (2015) they have been using Raspberry Pi for certain image processing modules in their project. We are using it as it is a small, handy computer, capable of doing high-end complex work and it can easily perform the processing of images used in facial recognition in our project.

The major difference between Eagle Eye and our developed project is that the methods and algorithm used are different and less tortuous. Eagle Eye uses its database to search the person whereas in our project the system searches for the person whose data has not been stored in the system database.

The working of Intelligent Surveillance System depends on two things, i.e hardware and software. The hardware part consists of an infrared CCTV camera along with Raspberry Pi 4 (Model B), an ethernet cable, USB Type-C Raspberry Pi Adapter, and a small buzzer alarm unit. The software part includes the code and certain face detection algorithms that we have developed according to our need for the working of the system which has been developed in Python 3. We are just implementing the algorithms to work in a faster and more precise way to get a high accuracy rate. The main work of this system is to use the database of trusted people for facial recognition to identify the genuine person and intruder based on the human face detected in the video surveillance zone by CCTV Camera. If any unauthorized person is detected in the surveillance zone then the administrator is informed with the help of notification over the text message and the burglar alarm is initiated. It uses the Viola–Jones algorithm to extract the Haar Cascade features for live detection of human faces. There will be one or two administrators responsible for surveillance system who will act accordingly when an intruder is detected by the CCTV camera. The system is connected to a LAN network over an ethernet cable while the system is on surveillance mode; it is connected to the Internet for sending the text message using Python API to the registered number of the administrator.

8.2 PROPOSED METHOD

8.2.1 RESEARCH DESIGN

This research has been designed with three main procedures to achieve its goal and serve the purpose for which it has been designed. They are as follows:

- Live face detection
- Face recognition
    i. Haar features extraction
    ii. Face matching from database
- Security protocol

The first step of the system is to detect the human face from the surveillance zone which is acquired from the CCTV camera. The second step is to recognize the faces detected in the CCTV camera, this step includes two main steps in which first step is to extract the Haar features using the Voila–Jones algorithm of the detected face and in the second step it matches those Haar features with the existing database of trained datasets. In the third step, if the detected face is matched with the existing database then there is no need to worry else it will start the security protocol by sending the text notification to the administrator and initiating the buzzer alarm.

In the face recognition process, the face features are extracted using the Voila–Jones algorithm to verify different faces on their facial characteristics and matched with trained dataset of registered person's face (Lienhart et al. 2002).

8.2.2 RELATED WORKS

Facial recognition is one of the most discussed and researched topics in machine learning as well as in computer vision area of application. The most popular algorithm in face recognition, the Viola–Jones (Viola and Jones 2001, pp. 137–54) face detection algorithm, uses basic facial characteristics to detect and recognize face. The face features based technique which is simpler and more effective is robust to position variations, the one who is more interested can read (Jebara 1996) for brief summary. The Viola–Jones algorithm feature includes various face detection methods. Many works have been done in the area of Intrusion Detection System to prevent intruders from entering particular places. In a project work (Ayi et al. 2017) we can see that they have used MATLAB for image processing to detect and track a particular face among various faces detected by the camera. It is an idea used for implementation of high security in video surveillance area to setup CCTV camera in such a manner that it can detect the motion of various person in a crowd after which it can detect and track down a particular person (Gerhard X. et al. 1996).

In the field of computer vision as well as pattern recognition we have seen that face recognition is the most well known topic in the last decades and many emerging companies has been working on it. In the early work of Turk and Pentland (Turk and Pentland 1991, pp. 586–91) we can see that they have done a classic work in the field of face recognition methods for the classification of face which has been done on the basis of Eigenfaces. It is previously stated (Sirovich and Kirby 1987, pp. 519–24) that face image can be shown as a linear combination of pictures i.e. Eigenfaces and its coefficients. In assurance of more security we need real time recognition of faces in video surveillance which will involve real time face recognition from the sequence of images captured from a CCTV camera (Jafri and Arabnia 2009, pp. 41–68). Many researches have been conducted in this field to improve computer vision in the area of real-time face recognition as well as to make the surveillance system intelligent which can work on their own intelligence. In many cases it has been seen that recognition is conducted on the basis of choosing certain good frames after which a suitable recognition method has been applied for intensity pictures to selected frames to find an individual (Chellappa et al. 1995, pp. 705–40).

8.3 IMPLEMENTATION

In our work we have focused on the identification of intruder and security measures that our system will take during video surveillance for which we have design various modules. The main feature of our Intelligent Surveillance System are as follows:

- Live face detection during video surveillance.
- Identify single or multiple faces that exists in system database.
- Track any unknown face i.e. intruder whose face is not stored in database.
- Start security measures if any intruder detected i.e. send text notification to administrator and turn on burglar alarm.

It is capable of detecting live faces from surveillance CCTV Camera using the Viola–Jones method (Shilpashree et al. 2015).

8.3.1 LIVE FACE DETECTION

The main goal of this phase is to accurately detect the human faces which will fall under the camera zone during video surveillance by a CCTV camera. The system has to determine in this phase whether any human faces are confronted in the surveillance area and if there are any human faces detected then the system must forward it to the second phase for the face matching and face recognition. For this purpose the Viola–Jones (Viola and Jones 2001, pp. 137–54) method comes in handy for face detection by the CCTV camera. The four major components in this algorithm are Haar features, integral image, adaboost, and cascading (Figure 8.1).

In this algorithm the features are rectangular in shape. The Voila–Jones algorithm uses a base window of size 24 × 24 to start evaluating features in any given image.

Hence the first module in the system has been used for real life face detection of the people who come in the surveillance area.

8.3.2 FACE RECOGNITION

The next step after live face detection from the surveillance video is to match that face with existing faces in a database and ensure whether the system has detected any intruder or not. If an intruder has been found in the surveillance area then the system must initiate the security measures. So for facial recognition and matching the faces that have been detected by the CCTV camera with the existing database, we have built a separate module which uses a trained dataset of trusted persons' faces and uses the Local Binary Pattern Histogram (LBPH) algorithm (Ahmed et al. 2018) for recognizing the detected human face.

The Local Binary Pattern Histogram has a Local Binary Pattern operator which uses Haar features of a human face that simplified the local special arrangement of a face image (Chen et al. 2006). The operator simply combines the Local Binary Pattern with a histogram of oriented gradients (HOG) descriptor that can represent the face images with a simple data vector (Keval et al. 2008). The Local Binary Pattern makes it possible for the system to the shape and textures in a digital image. The algorithm divides an entire digital image into small regions from which the Haar features can be easily extracted (McCaffery et al. 2018). Then these extracted features helps in comparing the images that are found in live face detection from surveillance video and the face images that are stored in an existing database (Lander et al. 2003). There is a trained dataset of trusted persons that has been created from a particular histogram derived from the extracted Haar features of a particular face which

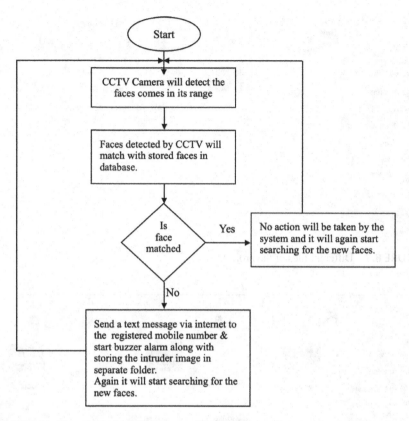

FIGURE 8.1   Flowchart of intelligent surveillance system using machine learning.

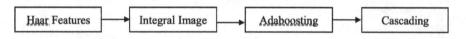

FIGURE 8.2   Flowchart of Viola–Jones algorithm.

will use a face recognition module for matching and recognition purposes (Padaruth et al. 2013). We have used a Local Binary Pattern Histogram algorithm as it is the most convenient and fast method for face recognition along with high accuracy rate (Hongshuai et al. 2017).

8.3.3 Security Measures

After fetching the data from face recognition module, the system will ensure that the detected person is an intruder or not. If the detected person is an intruder then the system will initiate security protocols to dodge back the intrusion. In V.D. et al. (2018) a model has been purposed to track down a particular person whose face is stored in database and when this person is found in the surveillance video then it uses

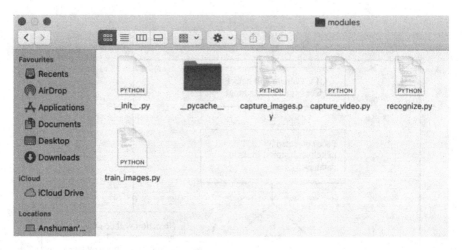

FIGURE 8.3   Different files of the project.

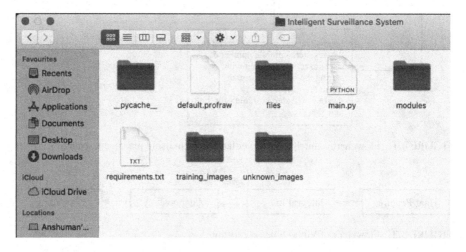

FIGURE 8.4   Different implementation modules.

text message notification using a GSM kit to inform where the particular person has last been seen by the CCTV camera; the message is sent to a particular target who has been tracking down that particular person, e.g.. crime branch, police department, or any other person. But in this project the system we will inform the administrator who has been responsible for the surveillance video recording or the owner of that place. Along with the text notification it will also start the burglar alarm to alert the nearby people.

In our system we have used Python API that will ensure to work precisely as soon as the system needs to send a text notification to a registered cellphone number. The system must be connected to the Internet till it is on surveillance mode using an

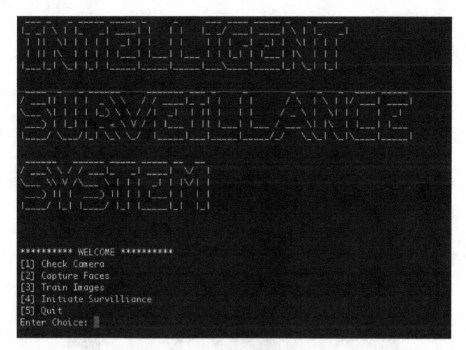

FIGURE 8.5   Interface menu of intelligent surveillance system using machine learning.

ethernet cable to ensure that API must serve its purpose of sending a text message to the registered cellphone number. When the system finds that the intruder has been detected in the surveillance video then it will use the API to send the text message to administrator, after which the system will start the burglar alarm to alert those nearby.

8.4   RESULTS

In this project every image of trusted persons' faces has a separate ID number in the database. Firstly we need to prepare the database of trusted persons' faces which will later be required by the system during face matching and face recognition. The system captures and stores 100–150 images of each person's face to train the dataset of the system database. The Local Binary Pattern operator extracts the Haar features from each face after then it uses it to identify and recognize the face information with the existing database. After which it simply prepares a data vector of each face that will be combined with a histogram pattern by the Local Binary Pattern that creates a trained dataset of each trusted person's face.

Then the system starts live detection of human faces during the surveillance video and captures each detected human face which will match with the existing dataset. It extracts the features of live detected faces and uses it for face matching. The system uses Local Binary Pattern Histogram algorithm for matching the human faces that

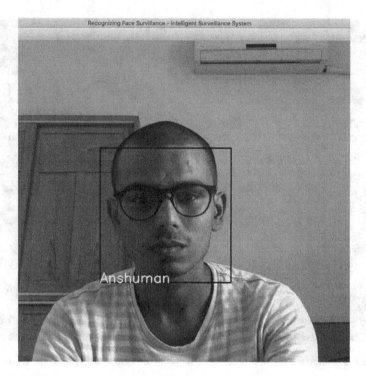

FIGURE 8.6   System has found a trusted face.

are live detected and then use the output data from the recognition module to justify whether the detected person is an intruder or not.

In Figure 8.6 it shows that when the system encounters a trusted person's face it then labels it with its name and ensure that the system is in a safe state.

In Figure 8.7 it shows that when a system found an unknown face that is not stored in existing database then it will label it as an intruder and initiate the security measures, i.e., sending text notification to the administrator and start the burglar alarm.

8.5   CONCLUSION

In this chapter we have described an intelligent system that focuses on intrusion detection with the help of trained dataset and alerts the administrator with text notification over cellphone along with a burglar alarm that is fixed with the system module. We used Raspberry Pi 4 (Model B) along with an infrared CCTV camera and a buzzer unit for the hardware components. We also need an ethernet cable for the Internet connection for sending text messages to the administrator. We have used Python API and certain Python libraries along with Open CV for facial recognition and face detection. Raspberry Pi 4 was found to be a compact and efficient computing processor for image processing with less power consumption and a low cost rate.

FIGURE 8.7 System found an intruder.

The system focuses on live face detection and face recognition from the surveillance video using a CCTV camera. It has always been a challenge to detect faces and recognize them in different light conditions with a high accuracy rate. We have found good results in maximum cases of various illumination conditions and image depths though in some conditions it has been found that the system encounters some errors in live face detection due to long distance of person from the CCTV camera and certain low light conditions. To tackle such problems we can connect more than one camera to the system for better results. Rest of all, it seems to work perfectly and gives a good accuracy rate of intrusion detection.

8.5.1 Future Enhancements

Although we have given our best efforts to develop this system, like all great inventions there are always some improvements that can make it even more better. It is found that in some cases of low lights and CCTV cameras installed at a far distance from the person that falls in the video surveillance zone, the system may find difficulty in detecting the face and hence may not recognize them. To overcome these problems in future we can implement a LiDAR (Light Detection and Ranging) sensor which is connected to the system. It will help in making the 3D representation of target which will ensure more accuracy in various low light conditions and high distant of the person from the CCTV camera.

REFERENCES

Aftab, A., J. Guo, F. Ali, F. Deeba, and A. Ahmed. 2018, May. LBPH Based Improved Face Recognition At Low Resolution. In *2018 International Conference on Artificial Intelligence and Big Data (ICAIBD)*.

Ayi, M., A, Kamal, Ganti, M. Adimulam, K. Badiganti, M. Banam, and G. V. Kumari. 2017, December. Face Tracking and Recognition Using MATLAB and Arduino. *International Journal for Research in Applied Science and Engineering Technology (IJRASET)*, 5(12).

Chellappa, R., C. L. Wilson, and S. Sirohey. 1995. Human and Machine Recognition of Faces: A Survey. *Proceedings of the IEEE*, 83: 705–40.

Chen, T., Y. Wotao, S. Z. Xiang, D. Comaniciu, and T. S. Huang. 2006. Total Variation Models for Variable Lighting Face Recognition. *IEEE Transactions on Pattern Analysis and Machine Intelligence*, 25(12): 1519–24.

EagleEye. 2017. China Surveillance System (Used by Chinese Police and Government).

Jafri, R., and H. R. Arabnia. 2009. A Survey of Face Recognition Techniques. *Journal of Information Processing Systems*, 5(2), 41–68.

Jebara, T. 1996, May. 3D Pose Estimation and Normalization for Face Recognition. Center for Intelligent Machines, Undergraduate Thesis, McGill University.

Keval, H., and A. Sasse. 2008. Can we ID from CCTV? Image Quality in Digital CCTV and Face Identification Performance. In S. S. Agaian and S. A. Jassim (Eds.), *Mobile Multimedia/Image Processing, Security and Applications, International Society for Optics and Photonics, Proceedings*, Vol. 6982, p. 69820K.

Kumar, V. D., S. Malathi, K. Vengatesan, and M. Ramakrishnan. 2018, July. Facial Recognition System for Suspect Identification Using a Surveillance Camera, *Journal of Pattern Recognition and Image Analysis*, 28(3).

Lander, K., and V. Bruce. 2003. The Role of Motion in Learning New Faces. *Visual Cognition*, 10(8): 897–912.

Lienhart, R., and J. Maydt. 2002. An Extended Set of Haar-like Features for Rapid Object Detection. In *International Conference on Image Processing, Institute of Electrical and Electronics Engineers (IEEE)*.

McCaffery, J. M., Robertson, D. J., Young, A. W., and Burton, A. M. 2018. Individual Differences in Face Identity Processing. Cognitive Research: Principles and Implications. 3(1), 21.

Owayjan, M., A. Dergham, G. Haber, N. Fakih, A. Hamoush, and E. Abdo. 2013, December). Face Recognition Security System. In *International Joint Conferences on Computer, Information, and Systems Sciences, and Engineering (CISSE 2013)*. University of Bridgeport.

Padaruth, S., Indiwarsingh, F., and Bhugun, N. 2013. A Unified Intrusion Alert System using Motion Detection and Faces Recognition. In *2nd International Conference on Mechine Learning and Computer Science (IMLCS), Kuala Lumpur*.

Ritter, G. X., and J. N. Wilson. 1996, January. *Handbook of Computer Vision Algorithms in Image Algebra*. CRC Press. LLC ISBN:0849326362.

Shilpashree, K. S., H. Lokesha, and H. Shivkumar. 2015, May. Implementation of Image Processing on Raspberry Pi. *International Journal of Advanced Research in Computer and Communication Engineering*, 4(5).

Sirovich, L., and M. Kirby. 1987. Low-Dimensional Procedure for the Characterization of Human Faces. *Journal of the Optical Society of America A*, 4(3): 519–24.

Turk, M. A. and A. Pentland. 1991. Face Recognition Using Eigenfaces. In IEEE Computer Society Conference on Computer Vision and Pattern Recognition. pp. 586–591.

Viola, P., and M. J. Jones. 2001. Robust Real Time Face Detection. *International Journal of Computer Vision*, 57(2): 137–54.

Zhang, H., Z. Q. Liping, and Y. G. Li. 2017. A Face Recognition Method Based on LBP Feature for CNN. In *IEEE 2nd Advanced Information Technology, Electronic and Automation Control Conference (IAEAC)*.

9 Cyber Security

An Approach to Secure IoT from Cyber Attacks Using Deep Learning

Sapna Juneja, Abhinav Juneja, Vikram Bali, and Hemant Upadhyay

CONTENTS

9.1 INTRODUCTION

Internet of Things (IoT) is a collection of devices that are connected with each other via some communication links and these devices can transfer data to each other through some wired or wireless connection. The IoT market is increasing day by day and this is expected to increase by 212 billion $ by the end of this year (Hutchison 2019).

IoT devices can be categorized into two types: Customer oriented and industry oriented (Al-Fuqaha et al. 2015). Customer oriented devices are the devices that can be used for domestic purposes e.g. washing machine, smartphone, smart TV, smart watch, CCTV cameras etc., whereas industry oriented IoT Devices are being used in smart cars e.g. healthcare, smart cities etc. Further the IoT market is expanding day by day and this ratio is increasing at a rate of 31 percent every year (Axelrod 2015). The growth in IoT industry can be represented by Table 9.1.

As the market of IoT industry is increasing, it is obvious that the numbers of IoT devices are also rapidly increasing in the whole world. So a large amount of voluminous

TABLE 9.1
Market value of IoT in billion$ from year 2016 to 2020 (Al-Fuqaha et al., 2015)

Category	2016	2017	2018	2020
Consumption	532515	725696	985348	1494466
Business Cross Industry	212069	280059	372989	567659
Business-Specific	634921	683817	736543	863662
Total	1379505	1689572	2094881	2925787

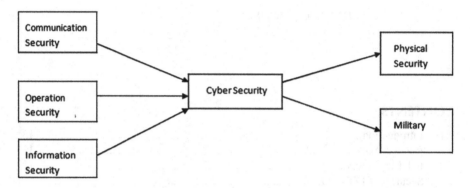

FIGURE 9.1 Domain of cyber security.

data has been transferred by these devices and there is always a threat of cyber attack on such data. Hence there is always a need to protect or secure these IoT devices from outside attacks so that data remains safe (Shukla 2018). The cyber security in IoT devices can be measured by different areas of security as shown in Figure 9.1.

In the domain of cyber security, various types of security measures are used to support different purposes and applications. Communication security is used to protect from unauthorized access to attackers in the communication and telecommunication and to deliver the content to actual receiver (Diro and Chilamkurti 2018). Operation security is the process of securing small pieces of data that can be provided by higher officials and decision-making bodies. Information security is used to protect any form of printed or documented data from intruders (Hou et al. 2019). Physical security is to protect any kind of hardware, software, or any other information that can be damaged or lost from some physical event (Al-Fuqaha et al. 2015). Usually the natural disasters can trigger such kind of physical events like floods, earthquakes, etc. Military security is carried to protect military data and networks from attackers and to save the confidential data of the country (Yu et al. 2017). So, all the mentioned security techniques are listed under cyber security and the overall goal of cyber security is to protect such data from attackers or hackers. The main purpose of this chapter is to define cyber security using IoT.

9.2 RELATED WORK

The Internet of Things has had a remarkable impact in a number of fields, and researchers across the globe have provided various perceptions and explored its uses (Díaz López et al. 2018) presented a thorough study to elaborate the security aspects of IoT devices by taking into consideration various possible security events, their accountability and attack surfaces. They proposed a relationship among these three mentioned terms to identify the possible causes of cyber attack. SIEM system had been used to identify any occurrence of security break (Rahman et al. 2018) tried to explain common attacks and vulnerabilities of most common used IoT device and defined a threat model in this research work. Various hardware assisted techniques to ensure the security of IoT device were also explained Abomhara and Køien (2015) presented challenges and threats in the security of IoT devices. The paper demonstrated that attacks from criminals and security agencies are very difficult to predict and handle instead of attacks from independent hacker. Nieto and Rios (2019) defined HFC technique for cybersecurity, and in order to validate the same, they elaborated and tested three different profiles using CPRM Modelling technique. He stated that HFC is a suitable approach for digital analysis but this approach will not work well if the number of parameters will be more. (Liu et al. 2017) stated that to verify that any particular IoT architecture of cyber security is reliable or not, some special measures are needed to be taken. One of the most important measures is contract decoupling in which different devices with different communication protocols are protected. Another important measure was scalability. IoT system must be easily scalable so that any device within the world can easily be connected with the structure. He also elaborated that there should be diversity in the testing techniques of the IoT device so that device can be tested through various angels.

9.3 IOT KEY COMPONENTS

The key components which enable any IoT device (as shown in Figure 9.2) to functions are given below (Abdur et al. 2017):

1. Sensors
2. Wired or wireless connection
3. Server for storage of data
4. Routers or switches for interconnection

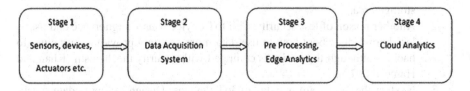

FIGURE 9.2 Layered Structure of IoT Device (Sikder et al. 2018).

The IoT must always be connected with a processor so that the process will control and monitor the data of that IoT device (Ray 2018).The sensors are being used to sense and collect the data from the particular IoT devices and then this data is stored by memory unit connected with the processor. User interface can also be provided for IoT device for the ease of use. Further all devices can be supported either by wired connection or by wireless connection (Sánchez et al. 2017). Further, the role of server in the IoT device is to run the application program and store the overall data of the device.

9.4 SECURITY OF IOT DEVICES

One important term that has been used in cyber attack is attack vector. An attack vector is the way by which any hacker can access any system or any network or server and can put any harmful information or virus into that system (Tiwari and Dwivedi, 2017). The hacker can also get the information of that attacked system. So the main reason behind the protection of IoT environment is to protect own information or device or property from others that may use that content without permission for any malicious purpose (Burhan et al. 2018). The second more important reason to protect IoT environment is that IoT devices are connected to the network in which many other devices may also be connected. So if one device is at risk, there may be a probability that other connected devices in that network are also at high risk of being hacked. Hence, it is quite important to prevent IoT devices or environment from malicious attacks. The way security is linked with IoT can be expressed in Figure 9.3.

As we move on from downward direction to upward direction in Figure 9.3, the lower two levels are used for simple Internet connection without any provided security measures. The third, fourth, and fifth levels are used to represent more level of security awareness The top three levels of the above figure are representing the maximum level of security awareness in the IoT environment.

If we think about the vulnerable points where chances of attacks are maximum, we will observe that both hardware and software are at equal risk of attacks by the attackers.

Security issues of IoT devices (Abdur et al. 2017) can be categorized as:

1. Manufacturers of IoT devices spend a lot of time on their applications according to the demands of the market but they usually don't focus on the security of IoT devices.
   Hence the devices become more prone to attacks (Díaz López et al. 2018).
   The security issues of the devices due to manufacturing are weak passwords, less secure hardware, no availability of security updates, and insecure storage area.
2. Another reason of less security with IoT device is user's ignorance and user's less knowledge about how to use data more securely providing less clues to hackers although the device is equipped with security mechanism (Khan and Hameed 2018).
3. Next in the queue are updates of IoT devices. During online update mechanism of any devices, data is usually transferred back by the device, though

FIGURE 9.3 Cyber security awareness triangle (Bada et al. 2014).

the network. This increases the vulnerability of hackers for getting the information about the device (Abbas et al. 2019).

Security of any IoT device also depends upon the location where this device is installed (Al-Mohannadi et al. 2016) e.g. the chances of vector attacks on camera that is located outside the home are more as compared to any wearable gadget of user.

9.5 POSSIBLE AVAILABLE ATTACK VECTORS ON IOT

There are various methods and techniques using which the attackers can hack the data by applying various attack vectors. The technique which they adopt to hack the data is as:

First they try to identify the target and then try to attack on that target using social media, email, chat, or some other means. After entering into the target user's network, they try to encode the network and apply various tools for breaking the security. After breaking the security, they usually insert the virus into the network or the system. Lastly, the attackers try to obtain the sensitive information out of the system to get the benefit from that data.

The various attack vectors and their impact on the IoT devices can be described as:

1. *Malware*: In case of IoT, the malware attack is somehow different from traditional malware. Malware is basically software that has been designed to get access of any system with the added purpose of damaging that system or its

information. Malware can only damage the systems that have been connected to the Internet. As the IoT devices are always connected with the Internet, hence they are at higher risks of getting attacked by malwares.

In order to protect IoT devices from malware attack, one should secure their network. This can be done by applying the firewall in the personal network. Further high security passwords can be applied on the routers.

2. *Keyloggers:* A keylogger is basically software that has been designed to track the data by the keys which anyone presses on his keyboard in such a way that user is unaware that something unusual is happening. Hackers can get very private information by this method e.g. ATM pin, passwords, etc. (Usman et al. 2017). For protecting the data from keylogger, some preventive measures should be taken such as: (a)
While opening an attachment in the mail, first confirm that the mail is received from actual user or some suspicious user (Usman et al. 2017). (b) User should be advised to change the passwords from time to time and select the option of two level securities for login (Buchanan et al. 2017). (c) Try to secure the data using antivirus or some other safety software (Singh et al. 2017).

3. *Phishing:* It is generally a type of cyber strike in which the attacker uses email to attack the device. The mail can be formed in such a way that the recipient believes that the message is from some authentic person and the content is also authentic. The mail usually appears with some attachment with it and upon downloading the attachment the security of the device leaks out (Samaila et al. 2018). In order to get the device and data secure from the phishing attack some measures should be taken as: (a) Before opening the attachment, first try to check the link (while avoiding clicking on the content) sent by the sender to verify whether there is something suspicious or not (Choi et al. 2016). (b) If there seems something doubtful, the receiver must contact the sender first in order to verify the email (Geetha et al. 2020). (c) Personal information should not be posted in social media as attackers can misuse such information in phishing mail to assure the recipient that the mail is authentic (Abdallah et al. 2018).

4. *Attack on IoT Sensors:* Nowadays, hackers try to attack devices, not only by attacking software but also attacking hardware such as sensors attached to the IoT. Hackers take over the device by attacking it and therefore access the data stored in it.To prevents the IoT sensor from attacks: IoT devices should be connected with a secure wi-fi network and the security protocol used should be based upon the latest security platforms. Another point is to regulate access to the IoT device to make it safe.

9.6 ENCRYPTION ALGORITHMS TO PROTECT DATA OF IOT DEVICES FROM CYBER ATTACKS

A few years back, the most common algorithms used to protect cyber data were Digital Signature Algorithm (DSA), Rivest–Shamir–Adleman (RES) etc. but as IoT devices nowadays demand more security, these algorithms are not sufficient to provide security to the same. So the algorithms that have been used to secure the IoT Devices are lightweight encryption algorithms. In the lightweight encryption algorithm, instead of applying encryption and decryption at the sender and receiver's end, the encryption algorithm can be applied to intermediate sensors and devices or to the processors that are processing the applications by keeping the process as lightweight as possible (Mulazzani and Sarcia 2011). The major benefits of using lightweight cryptography techniques are as Rumsfeld et al. (2016):

1. This lightweight technology provides more security, even at low power consumption and less number of resources (Archer, 2014).
2. Using lightweight process, more resources can be connected and hence can provide the security to more connected devices even using limited number of resources us (Park and Tyagi 2017).

Figure 9.4 shows the various lightweight cryptographic algorithms (Bugeja et al. 2018) and these can be categorized by different features as length of blocks, size of key, etc.

The size of blocks and keys in lightweight block ciphers remains the same. For encryption process in the block ciphers, two types of operations can be used: one is confusion and other one is diffusion. The process of confusion is used to relate ciphertext and encryption key whereas diffusion is used to convert ciphertext from ordinary to more sensitive for protection from any kinds of attacks (Patton et al. 2014).

A hash function has been used to protect the data and it takes input of any size and gives fixed size output. A lightweight hash function inhibits some properties for increasing the security to the data. These properties are collision resistance, pre-image resistance, and second pre-image resistance.

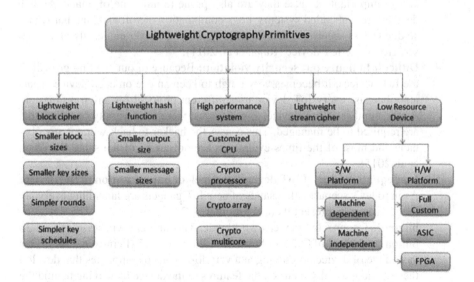

FIGURE 9.4 Lightweight cryptographic algorithms.

In lightweight stream cipher, every character of plain text is converted into cipher text. In this process of encryption, both the cipher key and the position of the plain text play an important role in the conversion (Ronen and Shamir 2016).

9.7 CHALLENGES IN CYBER SECURITY OF IOT DEVICES

Although various technologies and algorithms are available to secure the IoT devices, still the IoT industry is facing an everyday challenge for opting the same. The most common challenges that occur in IoT devices and IoT based applications are:

1. **Uncertain security standards of IoT:** The main cause behind the vulnerability of IoT devices is that various types of devices are generated by different developers in different organizations and hence each organization imposes its own security standard depending upon the features and usage of the IoT device. Hence no internationally applicable standard is used by any of the organization so it becomes quite easy for the attackers to attack on any particular device and break its security.

2. **Vital impact of IoT in human's life:** As nowadays everyone's life is influenced and dependent upon IoT devices ranging from smart healthcare, smart city, smart agriculture, smart transportation, etc. so due to a large amount of usage in everyday life, security standards applied are always high which usually attracts professional hackers to break out as they consider it as a challenge (Hameed and Khan 2018).

3. **Lack of security at hardware level:** It has always seen that developers always try to focus on the security of software alone and they usually ignore the security of the hardware as it is assumed that hardwares are safe from any kind of attacks. But it is not always true as the security of hardware device is also important because they are also prone to any type of attack. As IoT devices are embedded systems, i.e. a combination of software and hardware, so due to less security of the hardware part the attacker can easily attack the security of the IoT device (Rajan et al. 2011).

4. **Difficult to figure out security violation:** Because of outstanding growth of the IoT devices, it becomes very tough to keep an eye on each device separately because every IoT device uses different protocol for any transmission. As the number of IoT devices is enormous, there are a variety of things that are required to be managed. This causes the hacker to hack without any obstacle and most of the times even users are not aware of the same (Moosavi et al. 2015).

5. **Inadequate testing of IoT devices and lack of improvisation of the device:** As in today's scenario, thousands of new IoT gadgets are launching everyday in the market, at present there are more than 23 billion IoT connected devices worldwide. This amount will arise around 30 billion by the end of 2020 and over 60 billion by 2025 (Feinstein 2017). This vast range of devices is coming at a very low cost, so companies that develop these devices usually focus on the features of the device by leaving behind the security measures (Mohammadi et al. 2018).

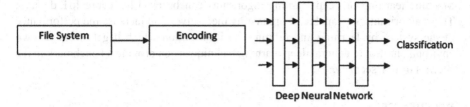

FIGURE 9.5   Deep learning model used to detect the cyber attack.

9.8   PROPOSED SOLUTION FOR ENSURING SECURITY OF IOT DEVICES

For Securing IoT systems and devices, the main approach is to monitor and analyze the complete system on continuous basis. So, the monitoring and analyzing can be performed using the deep learning approach. By applying deep learning, security breaches can easily be removed even with less number of computations (Colom et al., 2018). So the proposed methodology that can secure IoT environment from any kind of cyber attacks using machine learning or deep learning approach is as follows.

Deep learning is a technology based upon machine learning constructed using the neural network. Deep learning uses various channels of individual neural networks (Mahmud et al. 2018). So a type of neural network named as conventional neural network (CNN) is used to identify various objects that exist in the form of image. When input has been provided to these networks, the initial levels of the network are used to identify general characteristics from the input, while later stages identify exclusive characteristics.

So when any attempt to attack on the device is made, an application works in detecting the same. This application has a series of bytes with a fixed structure along with some instructions that executes on the given CNN architecture. These instructions are further encoded and applied onto the Deep Neural Network in order to obtain the classification. This Deep Neural Network has been trained by a dataset and is capable of identifying the normal program and any kind of attack. So when an input stream will be applied to these Deep Neural Networks, it will successfully separate out the normal program from the attack vector program.

9.9   CONCLUSION

As in today's scenario, IoT is rapidly increasing and it is much prone to possible attacks. So, some security measures should be considered during the development as well at the time of installation of IoT. In this research work, we have studied the possible available attack vectors and encryption algorithms that can be applied to protect IoT devices from cyber attacks. The major goal is to identify possible attacks and try to control them before they damage the system's data. If an organization is using IoT data, then it should focus upon the architecture of cyber security of the organization first, instead of focusing upon the performance or the cost of the device because if the security of the device will be higher, its performance will increase potentially.

Machine learning or deep learning algorithms can be used to secure IoT devices. These algorithms are capable of organizing themselves and increase the performance of the system by learning and refining the information which it gets from its own environment. So, in general deep learning techniques can provide a benchmark in the success of IoT security.

REFERENCES

Abbas, N., Asim, M., Tariq, N., Baker, T., and Abbas, S. 2019. A Mechanism for Securing IoT-Enabled Applications at the Fog Layer. *Journal of Sensor and Actuator Networks*, 8(1): 1–18. https://doi.org/10.3390/jsan8010016.

Abdallah, E. G., Zulkernine, M., and Hassanein, H. S. 2018. Preventing Unauthorized Access in Information Centric Networking. *Security and Privacy*, 1(4): e33. https://doi.org/10.1002/spy2.33.

Abdur, M., Habib, S., Ali, M., and Ullah, S. 2017. Security Issues in the Internet of Things (IoT): A Comprehensive Study. *International Journal of Advanced Computer Science and Applications*, 8(6). https://doi.org/10.14569/ijacsa.2017.080650

Abomhara, M., and Køien, G. M. 2015. Cyber Security and the Internet of Things: Vulnerabilities, Threats, Intruders and Attacks. *Journal of Cyber Security and Mobility*, 4(1): 65–88. https://doi.org/10.13052/jcsm2245-1439.414

Al-Fuqaha, A., Guizani, M., Mohammadi, M., Aledhari, M., and Ayyash, M. 2015. Internet of Things: A Survey on Enabling Technologies, Protocols, and Applications. *IEEE Communications Surveys and Tutorials*, 17(4): 2347–76. https://doi.org/10.1109/COMST.2015.2444095.

Al-Mohannadi, H., Mirza, Q., Namanya, A., Awan, I., Cullen, A., and Disso, J. 2016. Cyber-Attack Modeling Analysis Techniques: An Overview. *Proceedings – 2016 4th International Conference on Future Internet of Things and Cloud Workshops, W-FiCloud 2016*, 69–76. https://doi.org/10.1109/W-FiCloud.2016.29.

Archer, E. M. 2014. Crossing the Rubicon: Understanding Cyber Terrorism in the European Context. *European Legacy*, 19(5), 606–21. https://doi.org/10.1080/10848770.2014.943495.

Axelrod, C. W. 2015. Enforcing Security, Safety and Privacy for the Internet of Things. *2015 IEEE Long Island Systems, Applications and Technology Conference, LISAT 2015, January*, 1–6. https://doi.org/10.1109/LISAT.2015.7160214.

Bada, M., Sasse, A., Bada, M., Sasse, A., and Nurse, J. 2014. Cyber Security Awareness Campaigns: Why They Fail to Change Behavior. *International Conference on Cyber Security for Sustainable Society*, 11. www.cs.ox.ac.uk/publications/publication9343-abstract.html%0Ahttp://discovery.ucl.ac.uk/1468954/1/AwarenessCampaignsDraftWorkingPaper.pdf.

Buchanan, W. J., Li, S., and Asif, R. 2017. Lightweight Cryptography Methods. *Journal of Cyber Security Technology*, 1(3–4): 187–201. https://doi.org/10.1080/23742917.2017.1384917.

Bugeja, J., Jönsson, D., and Jacobsson, A. 2018. An Investigation of Vulnerabilities in Smart Connected Cameras. *2018 IEEE International Conference on Pervasive Computing and Communications Workshops, PerCom Workshops 2018*, 537–542. https://doi.org/10.1109/PERCOMW.2018.8480184.

Burhan, M., Rehman, R. A., Khan, B., and Kim, B. S. 2018. IoT Elements, Layered Architectures and Security Issues: A Comprehensive Survey. *Sensors (Switzerland)*, 18(9): 1–37. https://doi.org/10.3390/s18092796.

Choi, J., Seok, S., Seo, H., and Kim, H. 2016. A Fast ARX Model-Based Image Encryption Scheme. *Multimedia Tools and Applications*, 75(22): 14685–14706. https://doi.org/10.1007/s11042-016-3274-9.

Colom, J. F., Gil, D., Mora, H., Volckaert, B., and Jimeno, A. M. 2018. Scheduling Framework for Distributed Intrusion Detection Systems over Heterogeneous Network Architectures. *Journal of Network and Computer Applications*, 108: 76–86. https://doi.org/10.1016/j.jnca.2018.02.004.

Díaz López, D., Blanco Uribe, M., Santiago Cely, C., Vega Torres, A., Moreno Guataquira, N., Morón Castro, S., Nespoli, P., and Gómez Mármol, F. 2018. Shielding IoT against Cyber-Attacks: An Event-Based Approach Using SIEM. *Wireless Communications and Mobile Computing*, 2018. https://doi.org/10.1155/2018/3029638.

Diro, A. A., and Chilamkurti, N. 2018. Distributed Attack Detection Scheme Using Deep Learning Approach for Internet of Things. *Future Generation Computer Systems*, 82: 761–68. https://doi.org/10.1016/j.future.2017.08.043.

Feinstein, L. 2017. *IEEE Cloud Computing, January*, 26–33.

Geetha, R., Padmavathy, T., Thilagam, T., and Lallithasree, A. 2020. Tamilian Cryptography: An Efficient Hybrid Symmetric Key Encryption Algorithm. *Wireless Personal Communications*, 112(1): 21–36. https://doi.org/10.1007/s11277-019-07013-6.

Hameed, S., and Khan, H. A. 2018. SDN Based Collaborative Scheme for Mitigation of DDoS Attacks. *Future Internet*, 10(4). https://doi.org/10.3390/fi10030023.

Hou, J., Qu, L., and Shi, W. 2019. A Survey on Internet of Things Security from Data Perspectives. *Computer Networks*, 148, 295–306. https://doi.org/10.1016/j.comnet.2018.11.026.

Hutchison, D. 2019. *Survey High-Performance Modelling and Simulation for Big Data Applications* (Vol. 1, Issue 1). Springer International Publishing. https://doi.org/10.1007/978-3-030-16272-6.

Khan, F. I., and Hameed, S. 2018. Understanding Security Requirements and Challenges in Internet of Things (IoTs): A Review. *ArXiv, September*.

Liu, X., Zhao, M., Li, S., Zhang, F., and Trappe, W. 2017. A Security Framework for the Internet of Things in the Future Internet Architecture. *Future Internet*, 9(3): 1–28. https://doi.org/10.3390/fi9030027.

Mahmud, M., Member, S., Shamim Kaiser, M., Hussain, A., and Vassanelli, S. 2018. IEEE Transactions on Neural Networks and Learning Systems 1 Applications of Deep Learning and Reinforcement Learning to Biological Data. *IEEE Transactions on Neural Networks and Learning Systems*, 1: 1–17.

Mohammadi, M., Al-Fuqaha, A., Guizani, M., and Oh, J. S. 2018. Semisupervised Deep Reinforcement Learning in Support of IoT and Smart City Services. *IEEE Internet of Things Journal*, 5(2): 624–35. https://doi.org/10.1109/JIoT.2017.2712560.

Moosavi, S. R., Gia, T. N., Rahmani, A. M., Nigussie, E., Virtanen, S., Isoaho, J., and Tenhunen, H. 2015. SEA: A Secure and Efficient Authentication and Authorization Architecture for IoT-Based Healthcare Using Smart Gateways. *Procedia Computer Science*, 52(1): 452–9. https://doi.org/10.1016/j.procs.2015.05.013.

Mulazzani, F., and Sarcia, S. A. 2011. Cyber Security on Military Deployed Networks. *2011 3rd International Conference on Cyber Conflict, January 2011*, 1–15.

Nieto, A., and Rios, R. 2019. Cybersecurity Profiles Based on Human-Centric IoT Devices. *Human-Centric Computing and Information Sciences*, 9(1). https://doi.org/10.1186/s13673-019-0200-y.

Park, J., and Tyagi, A. 2017. Using Power Clues to Hack IoT Devices: The Power Side Channel Provides for Instruction-Level Disassembly. *IEEE Consumer Electronics Magazine*, 6(3): 92–102. https://doi.org/10.1109/MCE.2017.2684982.

Patton, M., Gross, E., Chinn, R., Forbis, S., Walker, L., and Chen, H. 2014. Uninvited
    Connections: A Study of Vulnerable Devices on the Internet of Things (IoT).
    *Proceedings – 2014 IEEE Joint Intelligence and Security Informatics Conference, JISIC
    2014*, 232–5. https://doi.org/10.1109/JISIC.2014.43.
Rahman, F., Farmani, M., Tehranipoor, M., and Jin, Y. 2018. Hardware-Assisted Cybersecurity
    for IoT Devices. *Proceedings – 2017 18th International Workshop on Microprocessor
    and SOC Test, Security and Verification, MTV
    2017*, 51–6. https://doi.org/10.1109/MTV.2017.16.
Rajan, M. A., Balamuralidhar, P., Chethan, K. P., and Swarnahpriyaah, M. 2011. A Self-
    Reconfigurable Sensor Network Management System for Internet of Things Paradigm.
    *2011 International Conference on Devices and Communications, ICDeCom 2011 –
    Proceedings*, 0–4. https://doi.org/10.1109/ICDECOM.2011.5738550.
Ray, P. P. 2018. A Survey on Internet of Things Architectures. *Journal of King Saud
    University – Computer and Information Sciences*, 30(3): 291–319. https://doi.org/
    10.1016/j.jksuci.2016.10.003.
Ronen, E., and Shamir, A. 2016. Extended Functionality Attacks on IoT Devices: The Case of
    Smart Lights. *Proceedings – 2016 IEEE European Symposium on Security and Privacy,
    EURO S and P 2016*, 3–12. https://doi.org/10.1109/EuroSP.2016.13.
Rumsfeld, J. S., Joynt, K. E., and Maddox, T. M. 2016. Big Data Analytics to Improve
    Cardiovascular Care: Promise and Challenges. *Nature Reviews Cardiology*, 13(6): 350–
    9. https://doi.org/10.1038/nrcardio.2016.42.
Samaila, M. G., Neto, M., Fernandes, D. A. B., Freire, M. M., and Inácio, P. R. M. 2018.
    Challenges of Securing Internet of Things Devices: A Survey. *Security and Privacy*,
    1(2): e20. https://doi.org/10.1002/spy2.20.
Sánchez, H., González-Contreras, C., Enrique Agudo, J., and Macías, M. 2017.
IoT and iTV for Interconnection, Monitoring, and Automation of Common Areas of Residents.
    Applied Sciences (Switzerland), 7(7). https://doi.org/10.3390/app7070696.
Shukla, P. 2018. ML-IDS: A Machine Learning Approach to Detect Wormhole Attacks
    in Internet of Things. *2017 Intelligent Systems Conference, IntelliSys 2017, 2018*
    (September), 234–40. https://doi.org/10.1109/IntelliSys.2017.8324298.
Sikder, A. K., Petracca, G., Aksu, H., Jaeger, T., and Uluagac, A. S. 2018. *A Survey on Sensor-
    based Threats to Internet-of-Things (IoT) Devices and Applications*. http://arxiv.org/abs/
    1802.02041.
Singh, S., Sharma, P. K., Moon, S. Y., and Park, J. H. 2017. Advanced Lightweight
    Encryption Algorithms for IoT Devices: Survey, Challenges and Solutions. *Journal of
    Ambient Intelligence and Humanized Computing*, 0(0): 1–18. https://doi.org/10.1007/
    s12652-017-0494-4.
Tiwari, V. K., and Dwivedi, R. 2017. Analysis of Cyber Attack Vectors. *Proceeding – IEEE
    International Conference on Computing, Communication and Automation, ICCCA
    2016, February*, 600–4. https://doi.org/10.1109/CCAA.2016.7813791.
Usman, M., Ahmed, I., Imran, M., Khan, S., and Ali, U. 2017. SIT: A Lightweight Encryption
    Algorithm for Secure Internet of Things. *International Journal of
Advanced Computer Science and Applications*, 8(1): 1–10. https://doi.org/10.14569/
    ijacsa.2017.080151.
Yu, W., Liang, F., He, X., Hatcher, W. G., Lu, C., Lin, J., and Yang, X. 2017. A Survey on the
    Edge Computing for the Internet of Things. *IEEE Access*, 6(c): 6900–19. https://doi.org/
    10.1109/ACCESS.2017.2778504.

10 Learning the Dynamic Change of User Interests from Noise Web Data

Julius Onyancha and Valentina Plekhanova

CONTENTS

10.1 INTRODUCTION

Current research work argues that user interest relies on the basis that the visiting time of a page is an indicator of the level of user interest on visited web pages (Ahmed et al. 2011). The amount of time spent in a set of web pages requested by the user within a single session reflects interest level of a user, on the other hand, high frequency of visit to a webpage signifies strong user interest (Wei et al. 2015). The page visit duration is determined by the difference between timestamp in a session, where sessions are created based on a static time-out period (Peng and Zhao 2010). Moreover, existing research works applies a standard threshold value during classification of web pages in relation to the degree of user interest. However, identifying web pages of user interest based on these measures face a number of challenges: (1) User interest change over time which means determining interestingness of a web page and assigning to a predefined class based on previous interest will impact the

147

quality of information in a web user profile. (2) The web is dynamic which means as web data evolves, the interest of a user changes thus affecting any web data classification based on standard threshold values. Considering these challenges, there is a need to learn user interests over time considering evolving data available on the web.

This chapter aims to demonstrate; (1) how evolving user interests impacts web user profiling process; (2) how dynamic threshold value can minimize loss of useful information during noise web data elimination process. The practical application of the proposed work will contribute towards learning noise in

web data prior to elimination, it will subsequently reduce the amount of useful information which existing tools identifies and eliminate as noisy. The rest of this chapter is divided into the following section: Section 10.2 evaluates the current research work on learning user interests. Section 10.3 presents the proposed work in addressing the defined problems, using illustrative examples. Preliminary experiments are introduced in Section 10.4 and finally the conclusion of this chapter.

10.2   EXISTING RESEARCH WORKS

Finding useful information from ever growing web is always a challenge due to increased data sources (Nanda et al. 2014). The time a user spends on a specific web page signifies the level of interest on the content available (Kabir et al. 2012; Wu et al. 2014). Further, the frequency of visits to a specific web content is an indicator of its interestingness to the user (Wei et al. 2015). For example, (Santra and Jayasudha 2012) used Naïve Bayesian classification model to identity useful information from noise considering time and frequency of visits. Malarvizhi and Sathiyabhama (2014) applied Weighted Association Rule Mining to find web pages visited by a user and assign weights based on interest level. The significance of a given page is determined by a weighted measure calculated by time and frequency of visits. The weighting of web pages depends on the number of requests and the time a user spend on the page.

Learning data available on the web in line with user needs is a key to finding what is useful and not to a specific user. Therefore, understanding how user needs

changes over time plays a more significant role to the process creating dynamic user profiles. According to Alphy and Prabakaran (2015) and Grčar et al. (2005), web user profiles are dynamic in relation to changes of user interests, it is these changes that prove difficult to identify and eliminate noise data from a web user profile. Such differences in user interests and preferences pose a challenge to existing tools applied in noise web data reduction process. In order to ensure only useful information is available to a user given the time of interest, it is necessary to learn their dynamic change to available web data. A dynamic approach to user interest learning considered in this chapter is influenced by various indicators such as time duration, frequency and the depth a user visits on a given web page. These measures will influence the interestingness of web data to a web user.

Valuable information about user interests on the web is hidden in user logs extracted from a web server. In order to ensure useful information is extracted from raw web log files, understanding user needs and how they vary is critical to this process. The outcome will provide a more dynamic approach to finding useful information that reflects changes in user interests. To find information that defines user needs from raw data, it is critical to the pre-processing of extracted data prior to profiling and subsequent learning process (Nithya and Sumathi 2012). Pre-processing eliminates noise data from web user logs (Begum and Vidyavathi 2016). The process involves building a web user profile through user, session, and page view identification in order to determine user interest level in relation to visited web pages. Han and Xia (2014) acknowledge that the web usage mining process can identify useful information from noisy data if the pre-processing of the web log takes into account the level of user interest.

Several machine learning tools have been proposed for extracting useful information from the noisy web. Santra and Jayasudha (2012) applied a Naïve Bayesian Classification (NB) algorithm to find information about visitors to a website captured in the form of web logs extracted from the website. The aim was to understand user needs based on interaction with the content available on the web. Prior to learning what users are interested in, irrelevant content such as advertisement banners, images, and screensavers were eliminated during pre-processing stage. A naïve Bayesian Classification model was then applied to distinguish usefulness and noise content based on time and frequency of visit measure. Sripriya and Samundeeswari (2012) proposed an algorithm based on neural networks to determine how many times a page appears in extracted logs of web data. The number of times a page appears in the logs signifies how frequent the page was visited by a user, hence its interest level. The weight of the page defined by time and frequency of visits is used as measure to evaluate the interestingness of a web page to a given user. Azad et al. (2014) applied kNN to web data logs to find useful information from noisy web log files; their focus was on local noise, for example, advertisements, banners, navigational links etc. Extracted data from the web server is interrogated to determine its source which is the initial stage of cleansing in order to

remove traffic from bots and other irrelevant logs (Malarvizhi and Sathiyabhama 2014) proposed a Weighted Association Rule Mining method to find content from web pages that is interesting of users. Similarly, the interestingness of a web page was determined using weighting approach. In their work, interestingness of a web page is determined by its weight which is defined by the page visits' duration and the number of times a user returns to the page over a specific time. Web pages with low weighting are considered irrelevant while those with high weighting are useful to a user (Gupta et al. 2016; Tyagi and Sharma 2012).

10.2.1 The Influence of User Interest on the Noise Web Data Reduction Process

Finding useful information based on user interest is challenging due to the increasing amount of data available on the web (Nanda et al. 2014). The interestingness of web data is dependent on a user and the interests of a user may change over time. Wu et al. (2015) acknowledge that user interest relies on the principle that the visiting time of a page is an indicator of the level of user interest. The amount of time spent on a set of pages requested by the user within a single session forms the aggregate interest of that user in that session.

Learning user interest can either be implicit or explicit (Nanda et al. 2014; Rao et al. 2017; Wei et al. 2015). The explicit nature involves users providing feedback on their experience and if they found what they were looking for. This can

be in the form of a survey or a feedback form embedded on the page visited. Implicit learning mainly focuses on user journeys captured by the clicks and visits to a specific web page. Unlike explicit learning, user interaction with the content on the web is captured thus making it easy to track change of behaviour over time (Kim and Chan 2005). Therefore, understanding what web users are looking for on the web as well as how their interests are likely to change over time is easily captured in their journeys captured in web log data and analytics.

To understand the usefulness of data available on the web to a user at any given time, it is important to understand the interest of each user. The change of interests impacts how useful the content is available to them and subsequently the process of determining which information is useful and is not. In this work, the focus is on learning how usefulness of the content available on the web is driven by what web users need at a given time. In summary, data available on the web is useful if it meets the need of the user and not its content structure.

10.2.2 Addressing Dynamic Change in User Interests

The existing research acknowledges that user interests tend to change over time (Jiang and Sha 2015). For example, the entire set of user interests can include interests that are relevant to a wedding and thereafter it may change to shopping for a new baby. Thus, if the user's interests were to change over time, the profile would reflect these changes by adding web pages to categories recently viewed and removing web pages from categories no longer found interesting. Despite the influence of time and frequency measures in learning the interestingness of web data, the proposed research points out some challenges associated with these measures. Firstly, it is recognized that the more time a user spends on a web page, the more interesting the page is (Tang et al. 2010). The amount of time a user spends on a web page varies from one user to another, mainly due to familiarity with the website and reading speeds. A user struggling to find information of interest may also take longer on a web page, or they may attend to other activities outside the page. Secondly, web pages visited within a session can either be auxiliary or content pages (Munk et al. 2015). Auxiliary pages help a user to find web pages that are of interest; they act as a visiting path to a user "destination." The frequency of this type of page will be high, but the duration of the visit will be low, which therefore suggests that frequency on its own may fail to determine the interestingness of a web page.

10.3 PROPOSED RESEARCH WORK

The proposed research argues that relying solely on frequency and time duration is inadequate in determining the user interest level of a web page. As a result, it is difficult to identify and eliminate noise web data based on duration and frequency of page visit. User interests in web data can change quickly, while others can change gradually over time. Therefore, learning changes in user interest aids in understanding how interesting the requested web pages are to a user. The following are some of the critical aspects examined in this chapter:

- *Last date of visit to a web page category:* even though a user may frequently request web pages from a specific category, how long is it since the last request?
- *Frequency of change in user interest:* it is important to understand how often user interests change over time. Therefore, defining change frequency for requested web page aids in looking into factors that trigger such changes. For example, if the requested information is seasonal.

10.3.1 RECENCY ADJUSTMENT MEASURE

The proposed research aims to explore the recency measure, which is considered critical to determining the interestingness of web data. Recency in web data mining is the time within which information on the web is considered relevant to the needs/interests of a user (Aly et al. 2013; Dong et al. 2010). In other words, recency is measured by the "freshness" of web data i.e., the time information on the web is interestingness to a specific user which is defined by how recent a user visit to the web page was.

Learning how frequently a user visits a web page allows website owners/developers to understand user interest level in order to ensure only relevant

information is suggested. It is equally important to examine how long it is since a user requested a specific page. The aim of this is to ascertain whether the user interest has changed over a time. The proposed research considers the recency measure to reveal how recently a web page was visited as compared to how frequently it is requested by a user.

Recency is defined as the time since the last user visit to a web page (Chakraborty et al. 2017). The best way to capture the interestingness of a web page based on its recency is by determining the number of days since the last occurrence of *kth* web page in the *jth* user profile. For example, if x number of days have passed since the last time a user visited the "home and living" web page category, then it might indicate that though the user was interested before, this may not be the case any longer; the user might have already acquired what he/she was looking for. Instead of setting a fixed time that determines the interestingness of a web page in a user profile, learning interestingness of a web page gradually over time is considered ideal. For instance, a user may have recently visited a specific category of a web page, but the frequency and time spent on the category is gradually decreasing. Even though website owners/developers will continue suggesting relevant information, the time will come when a user will no longer be interested, hence such information becomes noise.

The proposed research considers the interest forgetting function to determine the rate at which user interests change over time. (Hawalah and Fasli 2015) applied the interest forgetting function to remove data that is outdated from a user profile. A gradual forgetting function determines the weight of the web page category based on the time of its occurrence. The weight is dynamically adjusted each time a user visits the associated web page; thus, the interestingness of a web page category is measured based on its appearance in a user profile over time. The concept behind this function is that the interestingness of web data diminishes gradually with time. Therefore, the most recent web page visits in a user profile are considered more interesting than the old ones (Schwab et al. 2001)

To ensure gradual changes in user interests are managed, a half-life of interest denoted as hl, is defined. The half-life span of interest is the rate at which user interests change over a period (Suksawatchon et al. 2015; Tavakolian et al. 2012). The half-life span is key in learning the interestingness of web pages in a user profile over time, considering user interests as they change; it defines the average rate at which a web page within a user profile becomes noise. Existing research argues that user interest reduces to half in a week (Gu et al. 2014; Sugiyama et al. 2004). However, it is important to define a half-life value that reflects a significant change in user interests. In the proposed research, the recency measure based on time of visit is determined using Equation (10.1)

$$Rec_m^j = \frac{log2}{h_l} \cdot \left(td_0 - td_n\right) \qquad (10.1)$$

where Rec_m^j = the recency adjustment weight of mth category for the jth user, td_0 = the current date, td_n = the date of the last occurrence of kth web page in the jth user profile, and h_l user is the half-life (in days).

10.3.2 Dynamic Threshold Values

The objective of using the support threshold is to find a single point, which is used to determine the interestingness of a web page. Existing research works in the web usage mining process are based on either the standard/uniform support threshold or the dynamic support threshold (Ou et al. 2008) . The standard/uniform support threshold is static and thus does not consider key aspects, such as change of user interests during classification of web page based on user interest levels. User interest information collected explicitly usually relies on a standard threshold value to measure the level of user interest in requested web pages. Several critical issues arise when using standard threshold support. For instance, Hawalah and Fasli (2015) and Kavitha and Kalpana (n.d.) argue that the threshold support value remains the same and does not learn the change of user interests; where a low threshold value is assigned, it may lead to a high level of noisy web pages identified as useful and vice-versa.

To overcome challenges associated with uniform threshold, dynamic threshold support is considered in order to ensure changes in user interests are adequately addressed. Wei et al. (2015) and Ying et al. (2012) observe that the dynamic threshold support value is mainly used to determine a range of user activities in relation to various measures used to learn user interests, for example, time and frequency of a user visit to a web page. Based on these measures, it is possible to determine the interestingness of web data in relation to varying user interests when the threshold value is dynamically defined. Dynamic threshold support plays a critical role in ensuring dynamic changes of user interests are considered during the identification and subsequent removal of noise in web data. Instead of setting a standard threshold value based on the weight of a web page, the proposed research makes use of dynamic threshold value. Hawalah and Fasli (2015) proposed a mechanism to calculate a threshold value that reflects changes in users' browsing behaviour; their mechanism is based on the frequency of user interest in each web page category. The proposed research considers a similar approach to (Hawalah and Fasli 2015), but instead of using frequency, interest category weight, which incorporates both frequency and length of visit to a web page category, is used. This is to ensure that the interestingness of web pages in a user profile reflect the time and number of visits. Recency adjustment measure is considered mainly to learn changes in user interests. Subsequently, the process of identifying noise web data is managed considering changes in user interests. The following key issues are considered:

- A web user profile contains all web pages that are perceived as interesting to a given user, but whose degree of interest varies.
- Generally, useful information on the web is assigned high weight, unlike noise data, which is assigned low weight.
- The threshold support to determine the interestingness of a web page is defined based on the interest weight of a web page.

The recency adjustment measure is first defined, as in Equation (10.2), then the standard deviation (α) of *kth* web page for the *jth* user.

$$\alpha = \sqrt{\frac{1}{N}\sum_{i=1}^{K}\left(W_{m_i}^j - Rec_m^j\right)^2} \qquad (10.2)$$

where α is the standard deviation, K is the total number of web pages in i^{th} user session, *wmij* is the interest weight of mth category in i^{th} session for the jth user, and *Recmj* is the recency adjustment weight of mth category for the jth user.

The threshold value is defined using Equation (10.3)

$$Threshold = \alpha + \left(\frac{\sum_{i=1}^{K}W_{m_i}^j}{K}\right) \qquad (10.3)$$

where $\sum_{i=1}^{K}W_{m_i}^j$ is the interest weight of mth category in i^{th} session for the jth user.

The dynamic threshold value not only possesses the ability to manage a change of user interests but also ensures useful information is not lost as a result of the uniform threshold value. Given this, dynamic threshold support is key to determining user interest level in a visited web page prior to noise data elimination.

Therefore, to determine a dynamic support threshold value, the evolving nature of web data, as well as user interest, should be considered.

Some critical issues justify considering dynamic threshold measure in the noise web data reduction process. (1) A web page with lower frequency and time of visit will be considered irrelevant to a given user profile. Where a threshold value is set too high, interest pages with lower threshold values will not be found. On the other hand, where the threshold value is set low, a lot of irrelevant web pages will be considered useful. (2) Interestingness of information on the web varies. Seasonal web data tends to attract more attention than general information that is accessed daily. For example, the 2018 Football World Cup tournament will attract more traffic than the ongoing Brexit news. It is therefore important to understand the nature of web data and the interest a user express. This is because the threshold value set for these two types of information will vary.

Dynamic threshold values in web data classification play a critical role in ensuring dynamic changes in user interests are considered during the noise web data reduction process. Where a standard threshold value is applied in the web usage mining process, it will be difficult to obtain results that conform to a user change in interests. For example, a high threshold support value will yield less useful information and a low threshold support value will yield too many results. Dynamic threshold values are defined by learning previous user interest in a requested web page. As user interests change over time, the dynamic values also change. Second, it is important to acknowledge that noisy data can be potentially useful in future. User interests change and therefore so does the interestingness of web data. Using static thresholds will impact the quality of information available to a user given the fact that current interest data can be noise in future and vice-versa. In the proposed research, dynamic threshold values are used in the following scenarios: (1) implicit learning of user interests, i.e. when users do not directly reflect the interestingness of a web page, but instead their activities on the web determine the importance of a web page; (2) during classification of a web page, it is subject to change of interest.

10.3.3 User Interests Classification

Classification is one of the key processes in web usage mining and it has a significant impact on addressing problems with noise web data (Nanda et al. 2014). Classification takes an object and assigns a class label to it based on its attributes. Traditionally, classification aids in the creation of a user profile based on the user's level of interest in web pages requested. Ying et al. (2012) argue that user profiling is usually seen as a data classification problem because the interests of a user change over time. The objective of carrying out classification is to determine the target class of each record in a web log file based on varying user interests. Existing researches have proposed various algorithms to address data classification problems. For example, Santra and Jayasudha (2012) proposed an algorithm to find interested and not interested users.

Web page classification can be divided into binary or multiclass classification (Qi and Davison 2009; Waegeman et al. 2011). Binary classification defines data into two classes; based on user interest level, a web page can either be noise or useful. Multiple classification problems arise when data does not simply belong to one class, however. For example, a web page can be useful, noise or useful and noise. The proposed research recognizes that determining the interestingness of a web page as either interest or noise opens some critical issues in the web data reduction process. Given that user interest varies, and the web evolves, there is a

need to learn a web page considering user interest prior to determining its classification. When addressing classification problems, such as incorrect classification
of web pages due to varying user interests, a set of web pages and a class label
are provided. For example, if the weight of a web page requested by a user meets
a specified threshold, it will be considered interesting otherwise, it is noise. The
class label "interesting" and "noise" are specified to allow for the classification of
web pages.

Defining class labels: The objective of defining a class is to learn whether each
page visit meets the set criteria. For example, k^{th} web page in the jth user profile is
assigned to a class based on the level of user interest. A class is a label whose value
can be described based on varying levels of user interest in visited web pages. For
example, each page visit by a user can be of interest, potential noise or noise. For
each of the weighted web pages, a classifier is defined that reflects the interest levels
of a user in relation to the corresponding web pages, which is later used for the classification process. Even though all web pages in a user profile can be considered
useful with varying interest levels, not all are of interest. Moreover, the user is not
directly involved in determining the interestingness of a web page. This work, therefore, considers time and frequency of page visit, user visit depth, and frequency of a
web page category, to learn user interest levels.

Consider a class label $CL = (cl1, cl2, cln...clN)$ where N is the maximum
number of predefined classes. For illustrative purposes, the following classes are
considered: $cl1$ = interest class, $cl2$ = potential noise class and $cl3$ = noise class.

Interest Class: Web pages whose interest level meets the threshold value are
assigned to the interest class. To ensure the interestingness of a web page reflects
the user interest, the threshold value is dynamically defined to avoid low or high
threshold value, as in the case of standard or uniform threshold values.

Potential Noise Class: As the web evolves, new information emerges that a user
is likely to not have visited. With no interestingness identified, such web pages will
be identified as noise and subsequently eliminated. In order to avoid this, a potential
noise class is defined that will consider the interest category of a web page to learn
its interestingness.

Noise Class: Noise web pages are determined by the interestingness of a web page
considering all interest measures such as duration and frequency of user visits to a
web page. Given that the interests of a user change, what is noise today can be
useful a different time, and for this reason, dynamic threshold values protect
against loss of useful information, as well as ensuring minimize noise levels.

10.3.4 NOISE WEB DATA LEARNING: ITS SIGNIFICANCE TO WEB USAGE MINING

The noise web data learning approach identifies different types of web data and
determines their interestingness to a user by considering a number of measures,
i.e. time and frequency, depth of a user visit to a web page and interest category
of a web page prior to elimination. Unlike existing research works where noise in
web data is identified and eliminated based on the relationship with the main content, the proposed approach considers a web user a key character in determining the
interestingness of web data. This work argues that the importance of data on the web

is mainly dependent on what is interesting to a user over time. Therefore, web data can be irrelevant or noisy if it does not satisfy the interests of a user.

10.4 EXPERIMENTAL DESIGN

The objective of this experiment is to investigate whether the change of user interest influence identification and subsequent elimination of noise web data. To illustrate the significance of user's change of interest in noise web data learning process, the proposed research first defines a time over which interest of a user on the visited web page is learnt. The experiment conducted considered 90 days. This is to ensure that there is enough gap for learning any change of user interest over the specified period. A user profile will contain web pages visited by a user within the specified period. If the user interests were to change over time, the weight of a web page category will be adjusted accordingly. This process considers the length of time interestingness of a web page is considered for the learning process.

Preliminary experiments were conducted to demonstrate how each measure, i.e., frequency and length of a user visit to a web page category performed on this type of data. The 90 days extracted web user access logs data was divided into two parts. The first-month data was used to build a user profile and subsequently learn their interest on visited web pages. The second part of the data was used to learn changes to user interests in order to define the interestingness of visited web pages prior to noise elimination.

The experiment is based on recency adjustment measure defined by Equation (10.2). Recency adjustment measure which is based on a forgetting function examines how a change of user interest influences interestingness of web pages in a user profile. The function does not just eliminate web pages as the user stops visiting the page but learns the user's interest change. Changes to user interest affect the weighting of a web page thus its classification status, i.e., either interesting or noise. Therefore, it is important to ensure a more dynamic and flexible approach to determine the interestingness of the page in line with the change of user interests. The goal is to ensure that the process of identifying and eliminating noise in web data is dynamic enough to reflect the change of user interest.

10.4.1 DISCUSSION OF RESULTS

Figure 10.1 shows varying user interest on web page categories visited over a 90 day period. It can be observed that the interestingness of a web page varies with time, for example, user interests in office products declined in a week from week 48–52 but again increased after week 1. Subsequently, interest in home and living category decreased after week 52 with no visits for the next five weeks. When user interest level on a web page category is gradually decreasing, it signifies that a user is no longer interested in the category hence its noisiness. Therefore, its classification will be influenced by the threshold value Equation (10.3) as user interests level change. On the other hand, the weight of web page category a user has recently shown interest will gradually increase, ultimately the classification will also be affected. The advantage of using this measure is that interestingness of a web page is not only determined by the duration and frequency of a visit to a web page but also over a time interests

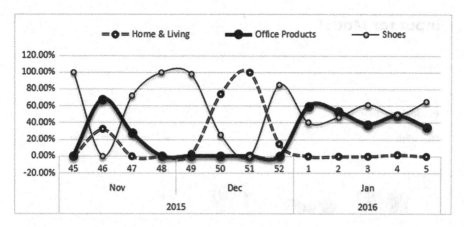

FIGURE 10.1 Dynamic change of user interest over 90 days' period.

of a user might change. Without considering the dynamic change of user interest, the amount of useful information otherwise eliminated a noise is likely to be high. Therefore, the choice of using dynamic measures to learn noise web data improves the quality of the information in a web user profile. As a result, the proposed measures support the claim that noise web data learning approach is user centric. This considers some factors which outperform existing machine learning tools applied in the noise web data reduction process.

Dynamic threshold value: The proposed research explores the impact of defining a threshold value that reflects the change of user interests. It is well recognized that interestingness of web page is dependent on the interest level of a user, which implies that the selection of a threshold value is a critical aspect in the process. For instance, if the threshold value is set low, then the output is likely to contain high noise levels. On the other hand, if the threshold is set high, then the chance of eliminating useful information is equally high. The dynamic threshold value applied by the proposed research considers the change of user interest over time as defined in Equation (10.3).

The results shown in Figure 10.1 are a classification of web pages which reflect the change of user interests. The process considers a dynamic threshold value which is determined by the interestingness of a web page over time. Results shown in Figure 10.2 are based on user visits within a week, we observe that the visited web pages are 45 percent interest, 34 percent potential noise and 21 percent noise. The proposed research acknowledges the existing research's viewpoint that user interest reduces to half in a week. However, it is important to consider such changes over time and this experiment consider interestingness of a web page over seven weeks.

The dynamic threshold value is determined based on recency adjustment measure. As time since last page visit increase, the interestingness of a web page

decreases thus becoming noise to a specific user. In Figure 10.3, the experimental results where user interest on visited web page is examined for seven weeks, it can be observed that the 78 percent increase in noise class is due to the time since a user expressed interest on web pages previously visited.

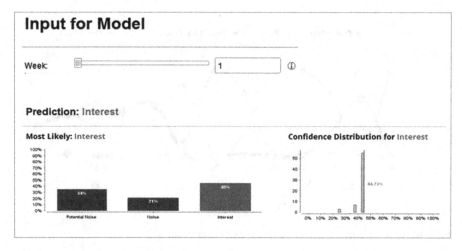

FIGURE 10.2   User interest level after one week.

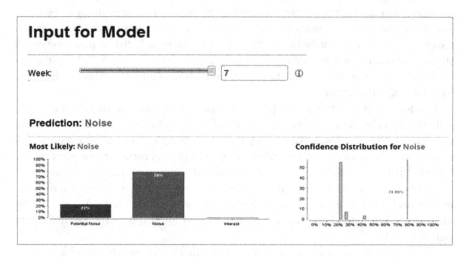

FIGURE 10.3   User interest level after seven weeks.

In summary, the experiment reveals that as user interest change over time, the interestingness of a web page is affected thus its classification. In essence, the level of nosiness in web data will be better managed if a change of user interest is considered prior to elimination. Therefore, the performance of the proposed work is measured by how well it defines and classify visited web pages in line with the level of interest.

10.5   CONCLUSION

The experimental results show a significant impact on finding useful information from the web when user interests are considered. One of the key aspects the proposed research consider critical is the dynamic change of user interests, the interest of a user on visited web pages is bound to change, but the process of managing such changes is challenging. The proposed research work through experiments conducted justifies the need to consider identification and elimination of noise web data as a user-centric approach because the importance of web data is better defined when user interests are considered.

REFERENCES

Ahmed, A., Low, Y., Aly, M., Josifovski, V., and Smola, A.J. 2011. Scalable Distributed Inference of Dynamic User Interests for Behavioral Targeting. In: *Proceedings of the 17th ACM SIGKDD International Conference on Knowledge Discovery and Data Mining, KDD'11. ACM, New York, NY, USA*, pp. 114–22. https://doi.org/10.1145/2020408.2020433.

Alphy, A., and Prabakaran, S., 2015. A Dynamic Recommender System for Improved Web Usage Mining and CRM Using Swarm Intelligence. *Sci. World J.* https://doi.org/10.1155/2015/193631.

Aly, M., Pandey, S., Josifovski, V., and Punera, K. 2013. Towards a Robust Modeling of Temporal Interest Change Patterns for Behavioral Targeting. In: *Proceedings of the 22nd International Conference on World Wide Web – WWW'13. Presented at the the 22nd international conference, ACM Press, Rio de Janeiro, Brazil*, pp. 71–82. https://doi.org/10.1145/2488388.2488396

Azad, H. K., Raj, R., Kumar, R., Ranjan, H., Abhishek, K., and Singh, M.P. 2014. Removal of Noisy Information in Web Pages. *ACM Int. Conf. Proceeding* Ser. 1.

Begum, H., and Vidyavathi, D. B. M. 2016. Web Recommender System using EM-NB Classifier 7, 3. *International Journal of Computer Applications* (0975–8887), 152(3): 9–12.

Chakraborty, A., Ghosh, S., Ganguly, N., and Gummadi, K. P. 2017. Optimizing the Recency-Relevancy Trade-off in Online News Recommendations. In: *Proceedings of the 26th International Conference on World Wide Web – WWW'17. Presented at the the 26th International Conference, ACM Press, Perth, Australia*, pp. 837–46. https://doi.org/10.1145/3038912.3052656.

Dong, A., Chang, Y., Zheng, Z., Mishne, G., Bai, J., Zhang, R., Buchner, K., Liao, C., and Diaz, F., 2010. Towards Recency Ranking in Web Search. In: *Proceedings of the Third ACM International Conference on Web Search and Data Mining – WSDM'10. Presented at the Third ACM International Conference, ACM Press, New York, USA*, p. 11. https://doi.org/10.1145/1718487.1718490

Grčar, M., Mladenič, D., and Grobelnik, M. 2005. User Profiling for Interest-Focused Browsing History. In: *Proceedings of the Workshop on End User Aspects of the Semantic Web*, pp. 99–109.

Gu, W., Dong, S., Zeng, Z., and He, J. 2014. An Effective News Recommendation Method for Microblog User. *Sci. World J.* https://doi.org/10.1155/2014/907515.

Gupta, R., Shah, A., Thakkar, A., and Makvana, K. 2016. A Survey on Various Web Page Ranking Algorithms, *International Journal of Advanced Computer Technology*, 5(1): 8.

Han, Y., and Xia, K. 2014. Data Preprocessing Method Based on User Characteristic of Interests for Web Log Mining. In: *2014 Fourth International Conference on Instrumentation and Measurement, Computer, Communication and Control. Presented at the 2014 Fourth International Conference on Instrumentation and Measurement, Computer, Communication and Control*, pp. 867–72. https://doi.org/10.1109/IMCCC.2014.182.

Hawalah, A., and Fasli, M. 2015. Dynamic User Profiles for Web Personalisation. *Expert Syst. Appl.* 42: 2547–69. https://doi.org/10.1016/j.eswa.2014.10.032.

Jiang, B., and Sha, Y. 2015. Modeling Temporal Dynamics of User Interests in Online Social Networks. *Procedia Comput. Sci.* 51: 503–12. https://doi.org/10.1016/j.procs.2015.05.275.

Kabir, S., Mudur, S. P., and Shiri, N. 2012. Capturing Browsing Interests of Users into Web Usage Profiles. In: *Workshops at the Twenty-Sixth AAAI Conference on Artificial Intelligence*.

Kavitha, D., and Kalpana, B. 2015. Dynamic Log Session Identification Using a Novel Incremental Learning Approach For Database Trace Logs. *International Journal of Scientific & Engineering Research*, 6(5) (May): 1175. ISSN 2229-5518.

Kim, H., and Chan, P. K. 2005. Implicit Indicators for Interesting Web Pages. https://dspace-test.lib.fit.edu/handle/11141/162.

Malarvizhi, S. P., and Sathiyabhama, B. 2014. Enhanced Reconfigurable Weighted Association Rule Mining for Frequent Patterns of Web Logs. *Int. J. Comput.* 13: 97–105.

Munk, M., Benko, L., Gangur, M., and Turcani, M. 2015. Influence of Ratio of Auxiliary Pages on the Pre-Processing Phase of Web Usage Mining. *E M Ekon. Manag.* 18: 144–59.

Nanda, A., Omanwar, R., and Deshpande, B. 2014. Implicitly Learning a User Interest Profile for Personalization of Web Search Using Collaborative Filtering. In: *2014 IEEE/WIC/ACM International Joint Conferences on Web Intelligence (WI) and Intelligent Agent Technologies (IAT)*. Presented at the 2014 IEEE/WIC/ACM International Joint Conferences on Web Intelligence (WI) and Intelligent Agent Technologies (IAT), pp. 54–62. https://doi.org/10.1109/WI-IAT.2014.80.

Nithya, P., and Sumathi, P. 2012. Novel Pre-Processing Technique for Web Log Mining by Removing Global Noise and Web Robots. In: *2012 National Conference on Computing and Communication Systems*. Presented at the 2012 National Conference on Computing and Communication Systems, pp. 1–5. https://doi.org/10.1109/NCCCS.2012.6412976.

Ou, J.-C., Lee, C.-H., and Chen, M.-S. 2008. Efficient Algorithms for Incremental Web Log Mining with Dynamic Thresholds. *VLDB J.* 17: 827–45. https://doi.org/10.1007/s00778-006-0043-9.

Peng, Z., and Zhao, M. S. 2010. Session Identification Algorithm for Web Log Mining. In *2010 International Conference on Management and Service Science*. Presented at the 2010 International Conference on Management and Service Science, pp. 1–4. https://doi.org/10.1109/ICMSS.2010.5576547.

Qi, X., and Davison, B. D. 2009. Web Page Classification: Features and Algorithms. *ACM Comput. Surv.* 41: 1–31. https://doi.org/10.1145/1459352.1459357.

Rao, K. S., Babu, D. A. R., and Krishnamurthy, D. M. 2017. Mining User Interests from User Search by Using Web Log Data. *J. Web Dev. Web Des.* 2.

Santra, A. K., and Jayasudha, S. 2012. Classification of Web Log Data to Identify Interested Users Using Naïve Bayesian Classification. *Int. J. Comput. Sci. Issues* 9: 381–7.

Schwab, I., Kobsa, A., and Koychev, I. 2001. Learning User Interests through Positive Examples Using Content Analysis and Collaborative Filtering. 30 2001. Internal Memo, GMD.

Sripriya, J., and Samundeeswari, E. 2012. Comparison of Neural Networks and Support Vector Machines using PCA and ICA for Feature Reduction. *Int. J. Comput. Appl.* 40, 31–6. https://doi.org/10.5120/5066-7434.

Sugiyama, K., Hatano, K., and Yoshikawa, M. 2004. Adaptive Web Search Based on User Profile Constructed without Any Effort from Users. In *Proceedings of the 13th Conference on World Wide Web – WWW'04*. ACM Press, New York, USA, p. 675. https://doi.org/10.1145/988672.988764.

Suksawatchon, U., Darapisut, S., and Suksawatchon, J. 2015. Incremental Session Based Collaborative Filtering with Forgetting Mechanisms. In: *2015 International Computer Science and Engineering Conference (ICSEC)*. Presented at the 2015 International

Computer Science and Engineering Conference (ICSEC), pp. 1–6. https://doi.org/ 10.1109/ICSEC.2015.7401418.

Tang, J., Yao, L., Zhang, D., and Zhang, J. 2010. A Combination Approach to Web User Profiling. *ACM Trans. Knowl. Discov. Data*, 5: 1–44. https://doi.org/10.1145/1870096.1870098.

Tavakolian, R., Beheshti, M. T. H., and Charkari, N. M. 2012. An Improved Recommender System Based on Forgetting Mechanism for User Interest-Drifting. *Int. J. Inf. Commun. Technol. Res.* 4: 69–77.

Tyagi, N., and Sharma, S. 2012. Weighted Page Rank Algorithm Based on Number of Visits of Links of Web Page 2, 6. *International Journal of Soft Computing and Engineering* (IJSCE), 2(3): 441–6.

Waegeman, W., Verwaeren, J., Slabbinck, B., and De Baets, B. 2011. Supervised Learning Algorithms for Multi-Class Classification Problems with Partial Class Memberships. *Fuzzy Sets Syst.*, Preference Modelling and Decision Analysis (*Selected Papers from EUROFUSE 2009*), 184: 106–25. https://doi.org/10.1016/j.fss.2010.11.012.

Wei, X., Wang, Y., Li, Z., Zou, T., and Yang, G. 2015. Mining Users Interest Navigation Patterns Using Improved Ant Colony Optimization. *Intell. Autom. Soft Comput.* 21: 445–54. https://doi.org/10.1080/10798587.2015.1015778.

Wu, S., Xiaonan, Z., and Yannan, D. 2015. A Collaborative Filtering Recommender System Integrated with Interest Drift Based on Forgetting Function. *Int. J. U- E-Serv. Sci. Technol.* 8: 247–64. https://doi.org/10.14257/ijunesst.2015.8.4.23.

Wu, X., Wang, P., and Liu, M. 2014. A Method of Mining User's Interest in Intelligent e-Learning. *International Conference Data Mining, Civil and Mechanical Engineering*, 74–6.

Ying, J.C., Chin, C.Y., and Tseng, V. S. 2012. Mining Web Navigation Patterns with Dynamic Thresholds for Navigation Prediction. In: *2012 IEEE International Conference on Granular Computing. Presented at the 2012 IEEE International Conference on Granular Computing*, pp. 614–19. https://doi.org/10.1109/GrC.2012.6468696.

11 Artificial Intelligence Techniques Based Routing Protocols in VANETs

A Review

*Mamata J. Sataraddi, Mahabaleshwar
S. Kakkasageri, and Sunilkumar S. Manvi*

CONTENTS

11.1 INTRODUCTION

Wireless network is an integrated part of today's lifestyle to provide ubiquitous connectivity. Ad-hoc network provides connectivity without any fixed infrastructure. The nodes in ad-hoc network are self-organized and self-maintained. MANET and VANET are the two types of ad-hoc networks differing from each other basically

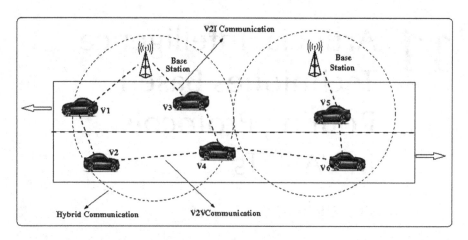

FIGURE 11.1 VANET architecture.

by the mobility of nodes (Cunha et al. 2016). Vehicles in VANET engage themselvs as both servers and clients to exchange information. As VANET is associated with life critical applications, there is need for understanding each aspect in architecture, features, applications, and issues of VANETs before implementing them. This application of Intelligent Transport Systems (ITS) should broadcast safety information to all the nodes with high level security.

The architecture of VANET (Chukwu and Agwu 2015) is as shown in Figure 11.1, where vehicles in VANET communicate with each other or through Road Side Units (RSU) to deliver the messages in time. VANET architecture can be categorized into three types, i) purely ad-hoc based also known as Vehicle-to-Vehicle (V2V), ii) infrastructure based known as Vehicle-to-Infrastructure (V2I), and iii) hybrid (V2V and V2I).

Research is growing in industry and academia to improve the VANET connectivity for autonomous driving. The key features of VANETs are as follows: highly dynamic topology, on-board sensors, unlimited battery, high mobility, large operational network, unlimited storage, and unlimited computing power (Li and Wang 2007; Wahid et al. 2018; Mezher and Igartua 2017; Wang et al. 2016).

To provide a variety of services to VANET users, vehicles are ready with smart portable device known as On Board Unit (OBU) with high computation and storage capability. VANET applications are categorized as safety and non-safety applications. Safety applications are related to increase safety on roads for drivers, passengers and pedestrians to avoid accidents which include delivery of emergency information, traffic monitoring, collision avoidance, road condition warning, lane change etc. (Sultan et al. 2014; Chen et al. 2010; Torrent-Moreno et al. 2009). The non-safety applications include driver and passenger comfortness (Moustafa and Zhang 2009; Shah et al. 2014).

Due to the high speed of vehicles, frequent network disconnection occurs which raise several issues in VANETs such as routing, link connectivity, network congestion, network management, information management, privacy and security etc. (Manvi and Kakkasageri 2008). The field of intelligent technology is autonomous in nature which has own computing capability and is interdisciplinary in nature, bridging computer science and engineering with its development of systems. Artificial intelligence uses these intelligent techniques to process and seek good solutions to the challenges occur in network such as routing. Some of the intelligent techniques include game theory, fuzzy techniques, bio inspired computational techniques, intelligent multi-agent systems, cognitive agents, neural networks, knowledge representation, etc. (Kenneth 2012; Sandeep and Medha 2012; Vidushi and Anoop 2013; Rahul et al. 2011).

Routing takes an open challenge in VANET, which is stated as choosing the best route for a message to reach its destination. Routing protocols are defined in order to communicate the nodes on VANET. Whenever the vehicles are not in the same communication range, different routing protocols are defined to deliver the data packets to their destination successfully. The main intention of efficient routing protocol is to minimize end to end delay, control overheads on the network which arises due to high mobility of vehicles, frequent link disconnection, and obstructions (Oubbati et al. 2015).

The chapter is organized as follows: Section 11.2 explains the importance of routing in VANETs and the concept of intelligent techniques used in VANET routing protocols. Various artificial intelligence techniques based routing protocols for

VANETs along with computational models are discussed in Section 11.3. Overview of artificial intelligence based VANET routing protocols are presented in Section 11.4. Section 11.5 concludes with future research scope in VANET routing protocols.

11.2 ROUTING IN VANETS

Routing is defined as selection of best route to deliver the data packets to the destination successfully. Much of research is going on to improve routing, considering various aspects and challenging features of VANETs. Due to frequent changes in network and variation in vehicle density different routing protocols are developed. Traditional routing techniques such as optimized link state protocol (OLSR) (Clausen and Jacquet 2003), Ad hoc On-demand Distance Vector (AODV) (Perkins et al. 2003) and Dynamic Source Routing (DSR) (Johnson et al. 2007) are not sufficient and fully applicable to VANET. To overcome these difficulties, ad-hoc routing protocols based on the information of vehicle's location, speed, and its direction of movement are presented in Tian et al. (2018), Alsharif and Shen (2017), and He et al. (2017). To improve delay and network overhead, different artificial intelligence techniques are applied to the conventional routing protocols of VANET.

11.2.1 WIRELESS STANDARDS FOR VANET ROUTING

To support applications of VANET, some standards for routing are proposed by different working groups which include Dedicated Short Range Communications (DSRC), Wireless Access in Vehicular Environments (WAVE) (IEEE 802.11p) and others. Telecommunication and automotive industry operators are allowed by these standards with some rules to create definite transportable routing devices (Salim and Abdelhamid 2014).

- **Dedicated Short Range Communications (DSRC):** DSRC (Sabouni and Hafez 2012) is a popular wireless standard developed to support ITS. It is used to connect different architectures of VANET to deliver messages for safety applications. According to DSRC, to implement wireless systems for Vehicle to Vehicle (V2V) and Vehicle to Infrastructure (V2I) communications, the US Federal Communications Commission (FCC) dedicated 70 MHz in the 5.850 to 5.925 GHz range. DSRC delivers data using a short-range radio at a rate of 27 Mbps which is considerably less price as compared to satellite or cellular communications (Morgan 2010).
- **Wireless Access in Vehicular Environments (WAVE):** Wireless Access in Vehicular Environment (WAVE) is main portion of DSRC that depicts the IEEE P1609.x standards focused on MAC and network layers to ensure protection in safety, health or environment and interference occurs from other devices or networks. WAVE is complex and is built over the IEEE 802.11 standards. The term DSRC is most commonly used compared to WAVE. On Board Unit (Mobile devices) and Road Side Unit (stationary devices) are radio devices defined in WAVE to offer or to utilize services.
- Vehicular communication is also attained by Bluetooth (IEEE 802.15), WiMAX (IEEE 802.16) and 4G Long-Term Evolution (LTE) standard. These transmissions can also be extended using new communication protocols like IPv6 (Salim and Abdelhamid 2014).

11.2.2 ISSUES IN ROUTING

Routing in VANET faces many issues while designing the routing protocol. Some of the issues are defined as follows (Salim and Abdelhamid 2014):

- **Scalability:** It is the most important issue in VANET routing which reduces communication performance and is defined as the capacity of a network to effectively handle more number of nodes. Scalability is better in rural areas due to low density where traditional routing protocols are well supported. For high dense VANETs such protocols are not suitable.
- **Computational complexity:** Due to complexity of network, a greater number of operations are performed in searching for a best path between source and destination to deliver the packets. For high dynamic network computational cost is more due to large execution time, and huge memory is required to maintain routing tables. Traditional routing protocols take a longer time to find the route due to rapid change in the network.
- **Self-organization and adaptability:** Autonomous routing, i.e., routing without human intervention, is self-organized routing. Traditional routing protocols require human intervention whenever changes in network occur.
- **Routing robustness:** The ability to ensure firm route between source and destination whenever link failure occurs.

11.2.3 ARTIFICIAL INTELLIGENT TECHNIQUES

Artificial intelligence is the theory of computer systems to develop and capable to perform tasks which normally require human intelligence, such as perception of vision, speech recognition, decision-forming, and language translation. Artificial Neural Network (ANN) is a group of statistical learning algorithms formed by the central nervous systems of animals, in particular the brain, and used to approximate functions which depend on more unknown inputs. Artificial intelligence includes a broad range of intelligent techniques (as depicted in Figure 11.2) (Eberhart and Shi 2007) which are as follows:

- **Game theory:** The study of strategic decision making. Game theory was originally developed by a German-born American economist – Oskar Morgenstern, and John von Neumann – a Hungarian born American mathematician. Specifically, it is "the study of mathematical models of conflict and cooperation between intelligent rational decision-makers." Whenever the actions of several agents are interdependent, game theoretic concepts apply. These agents may be individuals, groups, firms or any of these combinations.
- **Fuzzy logic:** A computing approach based on "degrees of truth" instead of basic "true or false." This idea was first invented in the 1960s by Dr. LotfiZadeh

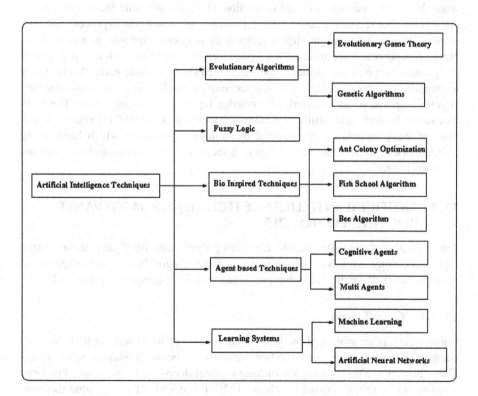

FIGURE 11.2 Classification of artificial intelligent techniques.

of the University of California at Berkeley. Fuzzy logic (FL) is an approach to computing based on "degrees of truth" using membership function.

- **Bio inspired computation:** The communication strategies and natural behavior of animal and even human biotic society are a motivation to design bio inspired algorithms to manage and control network related issues. This community includes behavior of colony of ants, swarm of bees, schools of fish, and even human society (Kromer and Musilek 2015).
- **Cognitive Agent:** Cognitive Agent is a software program that functions autonomously and continuously in environment. It supports the basic properties of an agent. To carry out actions intelligently and
- in flexible manner, to get immediate response for sudden changes in environment, learn from experience are some features of cognitive agent. The main components of an agent are Belief, Desire, and Intention (BDI), Belief, Obligations, Desire, and Intention (BOID), etc. (Kakkasageri et al. 2012; Hridya et al. 2016; Kakkasageri et al. 2017).

11.2.4 BENEFITS OF ARTIFICIAL INTELLIGENCE TECHNIQUES BASED VANET ROUTING

Traditional routing algorithms are not suitable for more dynamic, urban area network due to the issues mentioned in Section 11.2. To overcome these issues artificial intelligence technique based routing protocols are developed to process and seek good solutions. Artificial intelligent technology is interdisciplinary in nature, combining computer science information systems and engineering with its development of systems. Some of the artificial intelligent techniques include game theory, fuzzy techniques, bio inspired computational techniques, intelligent multi-agent systems, cognitive agents, neural networks, knowledge representation, and so on. The main motivation behind using artificial intelligence techniques for VANET routing is that most of these techniques are inspired by biological processes which have strong resemblance of natural communication of a species to the communication scenarios in networking.

11.3 ARTIFICIAL INTELLIGENCE TECHNIQUES BASED VANET ROUTING PROTOCOLS

This section explains different VANET routing algorithms using game theory, fuzzy logic, ant colony, schools of fish, bee swarm, genetic algorithm, cognitive agent, and neural network as intelligent techniques along with their computational model.

11.3.1 GAME THEORY

Game theory is an analytical method which uses set of mathematical tools to solve complex situation, that occurs when interactions between players takes place. These players are responsible for making rational decisions in the game. For large scale ad-hoc wireless networks such as MANET and VANET many game theoretic concepts were applied where the nodes are autonomous decision makers (Kapade and Deshmukh 2013). There are mainly two categories of game theoretic concept: non-cooperative (Lu et al. 2010) and cooperative game theory (Ketchpel 1993). The first concept deals with modeling competitive behavior, and cooperative game theory deals with cooperative behavior among a number of players and its impact on their decision-making process.

Whenever the source node in VANET starts sending its data towards its destination, it finds many paths using different QoS parameters. Due to rapid change in network, the data transmission takes up to some intermediate node, at which it has many options to select the next node to send data. In such situations, vehicle makes the decision by applying game theory based algorithm to calculate a matrix of all possible paths and to pick the best optimal path. Vehicles in the network need to cooperate with each other for safety applications. This situation can be represented using multi-player cooperative game theory (Wadea et al. 2017). This decision-making process at each intermediate vehicle of the path using multi-player cooperative game theory operates in the following sequence.

- **Network Coverage:** Here every vehicle needs to get information about all other vehicles in the network of communication. This is provided by RSU of their respective vehicle. Whenever a new vehicle joins the network, it computes the estimation of network coverage in each optional path using the previous and present coverage percentage of vehicle i.e., at time step k and at time step $k+1$ as given in Equation (11.1).

$$p(k+1|k) = f\left(p(k|k), z(k+1)\right) \qquad (11.1)$$

where $z(k)$ is the observation vector at time k, f is the updated function, and $p(k|k)$ signifies the vector of network coverage percentage at time step k that is computed using latest previous coverage percentage. When vehicle joins the network, the network coverage range is added to the observation vector. The utility vector for all paths is calculated at each vehicle and then the route with maximum utility is selected, which gives maximum network coverage.

- **Utility Function:** The utility value for all optimal routes is calculated at each vehicle by considering Travel Time Cost $\left(T_i^n\right)$, Time taken after reaching the route ($T_{c(i)}$), New network coverage percentage ($p_i^n(k+1)|k$), and are computed as follows:
  - **Travel time cost(T_i^n):** This is computed as given in Equation (11.2)

$$T_i^n = (L_i + S_d) * k \qquad (11.2)$$

where L_i is length of route i, S_d is Dijkstra least path length of the route i to the destination vehicle n and k is vehicle speed.

- **Network coverage time** $(T_{c(i)})$: This is to maximize the network coverage by computing the inter-vehicle distance based on the density of network and vehicle speed in the coverage area.
- **New network coverage** $(p_i^n(k+1)|k)$:This is expected by n^{th} vehiclein i^{th} route at time step k as cited in (11.1).

Now, the payoff cost of nth vehicle on route i at time step k is computed as given in Equation (11.3).

$$g\left(D(k), p_i^n(k+1|k), T_i^n(k), T_{c(i)}\right) = \frac{p_i^n(k+1|k)}{T_i^n(k) + T_{c(i)}} \qquad (11.3)$$

$D(k)$ – is the decision at time step k.

The utility of vehicle is calculated as the maximum of all routes starts with route i as given in Equation (11.4).

$$max_i\left[g\left(D(k), p_j^n(k), T_j^n(k), T_{c(i)}\right)\right] \qquad (11.4)$$

Each route j in the network is a subset of the route i, then the utility value of nth vehicle deciding on a route i is as given in Equation (11.5)

$$U_n^i = g\left(D(k), p_i^n(k+1|k), T_i^n(k), T_{c(i)}\right)$$
$$+ max_i\left[g\left(D(k), p_j^n(k), T_j^n(k), T_{c(i)}\right)\right] \qquad (11.5)$$

Now using Equation (11.6), each vehicle calculates its utility vector to all the routes as

$$U_n\left(d_1(k), d_2(k), \ldots, d_n(k)\right) \qquad (11.6)$$

Now the vehicle selects the route with the maximum utility value from the above vector for further decision.

- **Cooperation Scenario:** To avoid the selection of same route, the vehicles cooperation is required to know each other's decision. The cooperation factor h(D) is calculated using Equation (11.7).

$$h(D) = \frac{S_d^n(k) - T_i^n(k)}{\sum_{j=1, j\neq1}^{N} S_d^j(k) - T_i^j(k)} \qquad (11.7)$$

Where $S_d^n(k)$ - the minimum path is calculated using to Dijkstra's algorithm from vehicle $n's$ existing position to its destination. Hence for these vehicles, the utility function for every route is computed as shown in Equation (11.8).

$$U_n^i = g\left(D(k), p_i^n(k+1\mid k), T_i^n(k), T_{c(i)}\right)$$
$$+ max_i \left[g\left(D(k), p_j^n(k), T_j^n(k), T_{c(i)}\right)\right] * e^{h(D)} \tag{11.8}$$

The final best path is selected having maximum utility factor.

Similarly there are different routing protocols based on game theory approach for different application of the VANET. Time strategic game theory based on sequential bargaining routing scheme is proposed in Sungwook (2016). Using the vehicles cooperation the bidding scheme using cooperative game theory is explained in Muhammad et al. (2011). This selfish nature of vehicles bidding improves the welfare of the network by forwarding the message successfully. Stimulation of forwarding the message in VANETs using coalitional game approach is presented in Tingting et al. (2011). In Di et al. (2012), a routing protocol called Multi-Community Evolutionary Game Routing algorithm (MCEGR) is approached to overcome the problem of selfish behaviour of nodes in forwarding the messages using evolutionary theory. The work given in Muhammad et al. (2017) presents trust model for VANETs based on game theory. To maximize the signal-to-noise ratio (SNR) and capacity of channel in vehicle-to-vehicle (V2V) communication, a coalitional game theory based stable cluster formation in VANET is presented in Selo et al. (2019). Here SNR, network lifetime, and difference between speeds of vehicles are considered to formulate the value of coalition.

11.3.2 Fuzzy System

The things which are vague or not clear are referred as Fuzzy. In reality there are situations such as an event, function, or process that are continuously changing and are not defined as true or false always. Fuzzy logic gives flexibility for valuable reasoning of any imprecise and uncertain information. Fuzzy logic makes decision similar to the human. In digital logic system, 1.0 represents true value and 0.0 represents false value. But in fuzzy logic, intermediate values are also considered which are partly true and partly false.

The main four parts of fuzzy logic are:

- **Rule Base:** To govern the decision-making system, rule base is defined by the experts which is set of IF-THEN conditions and rules based on linguistic information.
- **Fuzzification:** The method of changing crisp values into fuzzy inputs is called fuzzification. The crisp values are the accurate values measured from sensors and are sent for processing.
- **Inference Engine:** This is the method of changing fuzzy inputs to fuzzy outputs using the set of fuzzy rules.
- **Defuzzification:** This is the method of changing the fuzzy outputs gained by inference engine into a crisp value using membership functions to reduce the error.

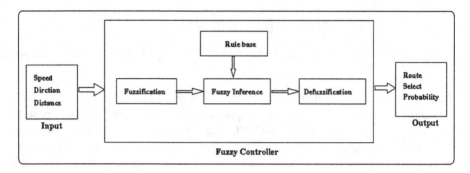

FIGURE 11.3 Fuzzy control system.

Consider the scenario of VANET for V2V communication, where each vehicle is moving in different direction with different mobility within the predefined area.

The best path between source and destination vehicle by the application of fuzzy logic method (Feyzi and Sattari-Naeini 2015) is as follows.

Ad hoc On demand Distance Vector (AODV) routing is the familiar routing algorithm in VANET which uses less number of hops to find the route. AODV reduces the network efficiency due to frequent changes in the network topology because of high speed nodes present in VANET. To increase the network efficiency three parameters viz speed, direction degree between two vehicles and distance of vehicles to destination have been considered to estimate the node weight using fuzzy logic. The fuzzy control system discussed in the above-mentioned research work is as shown in Figure 11.3.

- **Fuzzification:** The measured values (called crisp values) are converted in to linguistic values such as L-Low, M-Medium, and H-High, and are represented by fuzzy set. The crisp parameters considered are speed, direction degree between two vehicles, and distance of vehicles to destination.
- **Fuzzy rules and inference:** The fuzzy rules are derived by an expert that relates input and output of the fuzzy system to get optimum results. These rules are: IF premise, THEN consequence, where fuzzy input parameter is the premise and fuzzy output variable is consequence. These rules are used to find the weight of the node.
- **Defuzzification:** In this process fuzzy output values are combined and converted into crisp value.

The trapezoidal membership function is used for the inputs (speed, direction degree between two vehicles and distance of vehicles to destination) with three values as L-low, M-medium, and H-high as shown in Figure 11.4. The trapezoidal membership function for output (probability of selection) includes five values as very bad, bad, normal, good, perfect, as shown in Figure 11.5. The fuzzy set rules are given in Table 11.1.

In route discovery process the source node starts sending the Route REQuest packet (RREQ) to its neighbors and the neighbors also repeat the same process. This

TABLE 11.1
Fuzzy rules

Rule No.	Input			Output
	Speed	Distance	Direction	
1	L	L	L	Perfect
2	L	M	L	Good
3	L	M	H	Bad
4	L	M	M	Normal
5	L	H	L	Normal
6	M	L	L	Good
7	M	L	H	Bad
8	M	M	M	Normal
9	H	M	H	Very bad
10	H	H	H	Very bad

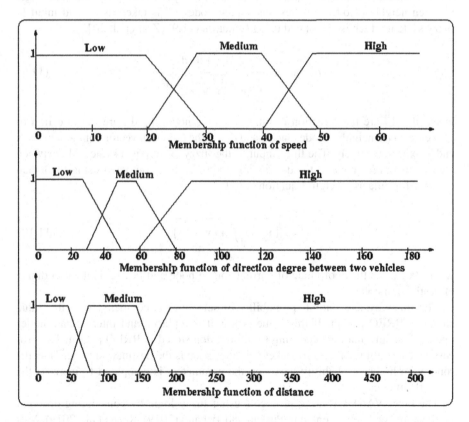

FIGURE 11.4 Membership function for speed, direction of degree, distance.

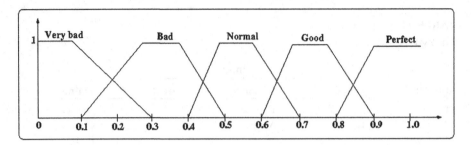

FIGURE 11.5   Membership function for probability of selection.

packet includes direction of the current source node, the speed of the packet receiver neighbor node, and the distance between current source node and destination node, assuming that every node knows the location of the destination node and the probability of selection which is initially considered one.

The neighbor node compares its own speed with the speed of previous source node. The node with higher speed is taken as first input to fuzzy system. The angle between neighbor node and previous source node "α" is taken as second input to fuzzy system which is calculated using Equation (11.9) (Zrar et al. 2013).

$$\alpha = \arccos \frac{d_{x1}d_{x2} + d_{y1}d_{y2}}{\sqrt{d_{x1}^2 + d_{y1}^2}\sqrt{d_{x2}^2 + d_{y2}^2}} \tag{11.9}$$

where d_{x1}, d_{y1} are the direction of current source node, d_{x2}, d_{y2} are the direction of packet receiver neighbor node, and d_x and d_y are the direction vector component on X and Y-axis respectively. The third input to the fuzzy system is distance "d" between the current source node to the destination node which is computed using Eucledian distance formula as given in Equation (11.10).

$$d = \sqrt{\left(x_2 - x_1\right)^2 + \left(y_2 - y_1\right)^2} \tag{11.10}$$

where (x_1, y_1) is the coordinate of current source node and (x_2, y_2) is the coordinate of destination node.

The fuzzy system output (probability of selection) is compared with the value stored in RREQ packet. If this value is less it is replaced and intermediate nodes update their information according to information stored in RREQ packet. The process repeats until the receiver packet neighbor node is the destination node. Thus the route with higher probability based on the parameters is computed to increase the network efficiency.

The other VANET routing protocols using fuzzy logic by considering different QoS parameters are presented in Purkait and Tripathi (2019), Rana et al. (2019), Miri and Tabatabaei (2020). In Purkait and Tripathi (2019) Packet Reception Probability with Changes of Intra Communication Distance (PRPwcICD), Node Speed

Difference (NSD), Link Expiration Time (LET) between source and neighbor node are considered as fuzzy inputs and Node Stability Cost (NSC) as the fuzzy output for multihop communication to forward the data packet. Both V2V and V2I scenarios are considered in this mechanism. In Rana et al. (2019), to select the next hop it considers next hop distance, closeness, speed of node, transmission rate of data, and direction of node movement as fuzzy input set to calculate weight factor of nodes for stable routing to transmit the data successfully to destination vehicle. In Miri and Tabatabaei (2020), the stable route is selected using fuzzy logic with vehicle mobility and available bandwidth as fuzzy input parameters to increase network utilization.

11.3.3 Bio Inspired Algorithms

This section explains different biological structured based routing algorithms in VANET. These algorithms are classified based on evolutionary process, swarm intelligence, and different biological inspiration source based routing.

- **Evolutionary algorithms (EAs)**: EAs are heuristic search algorithms based on Charles Darwin theory of evolution applied for different tasks (Daniel 2015). The main charecteristic of EAs is the existence of population measured by its fitness function which is degree of adaptation of an organism to environment. The bigger the fitness value, the organism is more fit. The different evolution mechanisms are DNA code, mutation, reproduction, natural selection, and genetic drift. Genetic algorithm (Xin-She 2014) is one computational technique evolved from evolutionary science proposed in 1970 which involves reproduction (crossover), natural selection, and mutation depending on fitness mechanisms to find best possible solution to the concerned problem as follows.
- **Reproduction (crossover)**: Reproduction takes place by randomly selecting two chromosomes and exchanging their genes to create an individual with greater fitness.
- **Mutation**: Mutation is used to maintain genetic diversity between chomosome generations. It is a unary operation based on probability of mutation which is usually very less (0.01) by making small changes at random to an individual gene. The different mutation types are one point mutation and uniform mutation (Salim and Abdelhamid 2014).
- **Natural selection**: The individual with high fitness is selected to pass on to the next generation.

Most of the applications of genetic algorithms in VANET are for route selection and cluster head selection (Guoan et al. 2018).

- **Swarm intelligence algorithms:** The algorithms which are inspired based on the interaction between living organisms. This includes ant colony optimization, fish school algorithm, bee algorithm, etc.
  - **Ant colony optimization:** This method was first invented by Dorigo et al., and is used to get the route between source and destination in VANET (Salim and Abdelhamid 2014; Kakkasageri 2019). This method is based on

how the ant colonies find the shortest route between their nest and food. This is done by their usual charecteristic of utilizing explosive substance called pheromones. This substance is perceived by them through their antennas which act as receivers. When they move on the ground they leave these pheromones and they are used as trail by other ants in the colony. If there is more than one food concentration, they choose the path with more food concentration. The ACO algorithm works as follows.

- Find number of ants Q. each ant finds its own route from source node i to destination node j, with the decision probability (given in Equation 11.11) made at each nodes in the route.

$$P_{ij}^k = \frac{\left[T_{ij}(t)\right]^\alpha \left[x_{ij}\right]^\beta}{\sum_{h \in Q}\left[T_{ih}(t)\right]^\alpha \left[x_{ij}\right]^\beta} \qquad (11.11)$$

where α and β are constants, $x_{ij} = \dfrac{1}{d_{ij}}$ is the quality of the link, and d_{ij}– distance between source vehicle i and destination vehicle j. $T_{ij}(t)$– pheromone quantity between vehicles i and j at iteration t.

- If the path selected is better than the current path, the selected path becomes current path and is stored in memory. Whenever an ant travels the quantity of pheromone at iteration t is calculated using Equation (11.12).

$$T_{ij}(t+1) = (1-P)T_{ij}(t) + \sum_{k=1}^{Q}\Delta T_{ij}^k(t) \qquad (11.12)$$

where P is coefficient of pheromone decay, Q is number of ants, and $\Delta T_{ij}^k(t)$ is the residual amount of pheromone by ant k at iteration t (Equation 11.13).

$$\Delta T_{ij}^k(t) = \frac{A}{L^k(t)} \qquad (11.13)$$

where A is constant and L is ant k's path quality at iteration t.

- **Fish School algorithm:** Fish naturally live in a group called a school to guarantee their existence and to avoid dangers. Every fish tries to move to the center whenever there are less fish in a group. When fish find food in their visibility area, they step toward the food concentration and the nearby fish will follow and get the food fast. The fish jump to the position where they gets more food concentration. This behavior of fish is used for VANET routing (Sataraddi et al. 2017) which operates in three phases viz. (i) vehicles grouping, (ii) identifying source and destination, and (iii) discovery of route. The route discovery is based on prey behavior of fish where the intermediate nodes are found by random position of node depending on the speed and distance between nodes using Equation (11.14)).

$$X_i(t+1) = X_i(t) + \left[X_j - X_i(t) \right] / \left\| X_j - X_i(t) \right\| * \text{Step} * \text{random}() \quad (11.14)$$

where, Xi (t+1) = x-axis coordinate of new neighbor node, Xi(t) and Xj = x-axis coordinate of source and destination, ‖x‖ = Euclidean distance.

- **Bee algorithm:** The food foraging behavior of bee (Janusz and Witold 2015; Nicolas et al. 2018) is used to find the route between two nodes. Food foraging is the behavior of bees, when bees look for new site for food. Bees are grouped as employee, observer, and inspect bees. If food source is found bee employee reside at the food source and observer hunt near the source. Then, inspect bees travel around the whole area. If there is no extra food source, the employee bee shifts to inspecting mode.

11.3.4 LEARNING SYSTEMS (LS)

LS are an artificial intelligence tool capable of learning through previous knowledge. It begins with the basic knowledge and is then adapt autonomously by learning, to increase its performance. LS includes machine learning and Artificial Neural Networks.

- **Machine learning:** Machine learning is a subset of artificial intelligence which offers the capacity of automatic learning to increase through experience without being clearly programmed. Machine learning is classified as follows:
  - **Supervised learning:** In this learning, the machine will be trained with labeled data. Regression and classification come under supervised learning, e.g., weather forecasting.
  - **Unsupervised learning:** Here the machine is trained with unlabeled data without any guidance. It is applied where the data is imprecise and uncertain, e.g., mining. It is categorized as an association rule and clustering rule.
  - **Reinforcement learning:** This is an autonomous (self-learning) in nature where an agent intermingles with its environment by using actions and discovers errors or rewards (Sutton and Barto 2018). Markov Decision Process and Q learning are two models of reinforcement learning.

Since VANET is imprecise and uncertain in nature, Q-learning as reinforcement learning is used for route decision. The Q value is defined based on the stability and continuity of neighbor nodes (Yanglong et al. 2019). Every node in the network maintains the Q(c, x) table whose values ranging between 0 and 1, where c and x are current and its neighbor nodes respectively.

- The Stability Factor (SF): The stability of the neighboring nodes is computed by its stability factor and is stated as the change in relative distance between neighboring times (given in Equation 11.15)

$$SF_t(c,x) = \begin{cases} 1 - \dfrac{|D_t(c,x) - D_{t-1}(c,x)|}{R}, & |D_t(c,x) - D_{t-1}(c,x)| \le R \\ 0, & Otherwise \end{cases} \qquad (11.15)$$

Where D (*) is the Euclidean distance between nodes and R is the reinforcement reward.

- The continuity factor (CF): This is defined as degree of nodes based on the number of nodes present in communication range of source node. The greater the degree of node the better connectivity resulting in stable routing. The continuity factor is computed as given in Equation (11.16).

$$CF(c,x) = \frac{NUM_x}{NUM_{max}} \qquad (11.16)$$

Where NUM_x is the degree of node x, which is in the communication range of present forwarding node c, and NUM_{max} denotes the upper limit of the node degree that the node can have.

- The reinforcement reward R and Q are updated using Equations (11.17) and (11.18) based on SF and CF of the neighbor nodes.

$$R(c,x) = CF(c,x) * SF(c,x) \qquad (11.17)$$

$$Q_c(c,x) = \alpha * \{R(c,x) + \gamma * Q_x(x,n)\} + (1-\alpha) * Q_c(c,x) \qquad (11.18)$$

- The distance relationship between c and x is given by distance factor which is computed by Equation (11.19).

$$DF_t(c,x) = \begin{cases} \cos\theta * e^{\frac{D_t(c,d) - D_t(x,d)}{R}}, & D(c,x) < R \\ 0, & Otherwise \end{cases} \qquad (11.19)$$

where θ is the angle between node c, node x, and destination node d.

- The next hop in the route is selected using the discount Q value computed as given in Equation (11.20).

$$Nexthop = arg \max_{x \in N_c} Q^d \left(d, x \right) \tag{11.20}$$

Where N_c is neighbor set of node c, Q^d is discounted value of Q given by Equation (11.21)

$$Q^d \left(c, x \right) = Q \left(c, x \right) * DF \left(c, x \right) \tag{11.21}$$

Other routing algorithms in VANET using machine learning are described in Yujie et al. (2019), Wei et al. (2015), and Zoubir (2019).

- **Artificial Neural Networks (ANN):** Artificial Neural Network is a computational network which works on studies of the brain and nervous system of human. ANN has hundreds of neurons called processing units or nodes configured for a specific application through supervised learning algorithms such as multi layer perceptrons (MLPs) with back-propagation learning algorithms. It comprises three layers: input, hidden, and output. Each node has a transfer function, weight, and an output. The node's weights are adjusted during training process to minimize the error and the network reaches the specified level of accuracy (Dasagoudar and Manvi 2008).

In VANETs, while cluster formation, vehicles misbehave by crossing the speed limits which leads to network disconnection. Hence there is a necessity of detecting the misbehaving vehicles. VANET using Quality of Service-Optimized Link State (QoS-OLSR) routing algorithm (Khatib et al. 2015) solves the issue of misbehaving vehicles in VANET. The protocol uses supportive detection scheme using artificial neural network (ANN), that aggregate judgments to avoid the independent results and benefit from the previous recognition using constant learning.

11.3.5 SOFTWARE AGENT SYSTEMS

An agent is software that operates at the nodes to perform a particular task (or tasks) independently and intelligently, even in changing of network conditions. Mainly there are two types of agent based routing approaches for VANET routing: cognitive agent and multiagent.

- **Cognitive agent based routing:** Human brain computation in artificial intelligence machines is not possible to follow 100 percent. Hence an agent has to be designed which acts and think like human brain is called cognitive agent. Cognitive agents are static agents that need more computation and databases. Cognitive agent architecture (Kakkasageri et al. 2012) shown in Figure 11.6 includes different static and dynamic intelligent agents. The different architecture models of cognitive agents are Belief Desire Intention (BDI), Belief Obligations Intentions Desire (BOlD), Cougaar and MicroPsi.

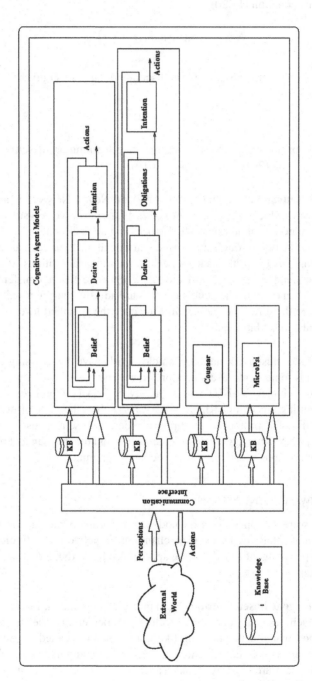

FIGURE 11.6 Cognitive agent architecture.

Dynamic routing scheme in VANET using BDI agent (Sataraddi and Kakkasageri 2017) is as shown in Figure 11.7 and operates as follows: (1) gathering of end-to-end delay and available bandwidth of each vehicle; (2) beliefs generation based on the end-to-end delay and available bandwidth; (3) development of the desire using generated beliefs; (4) route selection when desire is reached and intention is executed; and (5) route reselection when intention is not executed.

- **Multiagent based routing:** Multiagent systems are used for distributed and dynamic environments taking into cooperation, autonomy, distribution, and intelligence of agents. Each agent achieves its task in accordance with the additional agents. Multiagent based routing in VANETs (Samira et al. 2013) is as shown in Figure 11.8 and works as follows: (1) collect the context information of vehicles (vehicle's movement direction, interests, communication environments, etc.) using context agent (2) use different agents (cluster head agent, optimization agent) to minimize communication and to reduce network traffic, and (3) use router agent to search the best routes.

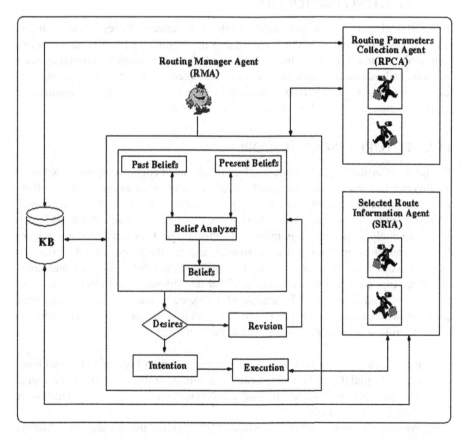

FIGURE 11.7   BDI agent based routing.

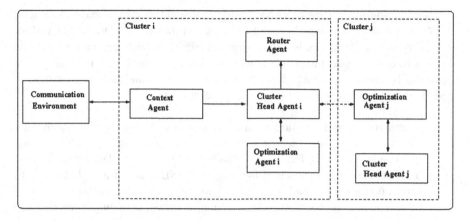

FIGURE 11.8 Multiagent System Architecture.

11.4 OVERVIEW OF ARTIFICIAL INTELLIGENCE BASED VANET ROUTING PROTOCOLS

Artificial intelligence techniques are more interest to researchers because of its ability to solve key problems in VANETs. Artificial intelligence techniques imitate more commonly human thought. It has comparatively low computation complexity, easy to apply, and maintain reasonable accuracy. However, several problems in applying artificial intelligence in VANETs remain open. Overview of the artificial intelligence techniques based routing algorithms are listed in Table 11.2.

11.5 FUTURE RESEARCH SCOPE

The future of artificial intelligence involves in advanced cognitive systems. Artificial intelligence technique will intelligently interact with human experts, offering them with clear explanations and solutions, even at the edge of the network. Artificial intelligence provides a more personalized experience to their customers. Artificial intelligence can analyze vast data way more efficiently. VANET with artificial intelligence can acquire information from diverse sources, expand the perception of driver, and guess to avoid possible accidents, thus increasing the comfort, efficiency and safety of the driving. In recent years, many artificial intelligence based routing protocols have recommended for VANETs because of their specific characteristics (Wang et al. 2018). However to deploy effectively artificial intelligence techniques for routing in VANETs, following research issues need to be addressed.

- Prediction of traffic flow is a big challenge in terms of optimization and modeling. Artificial intelligence predicts the minute traffic flow from the earlier records. However, some artificial intelligence methods and simulations are more time consuming.
- Practical implementations of artificial intelligence using congestion control methods need RSUs that are equipped with Graphics

TABLE 11.2
Overview of artificial intelligence techniques based routing algorithms

Research Works	AI techniques used	Architecture used	Computational complexity	Advantages	Disadvantages	Applications
CAV (Wadea et al. 2017)	Cooperative game theory	V2V	Medium	Improves network coverage	Increases travel distance	Avoids traffic congestion
TBOR (Sungwook 2016)	Distributed game theory	V2V	Low	High throughput	For high traffic, more packet delay occurs	Emergency message dissemination (accident, warning, etc.)
CBS (Muhammad et al. 2011)	Cooperative game theory	V2V	Medium	Guaranteed message transfer	Message generation and forward rate is more	Trust related applications
MCEGR (Di et al. 2012)	Evolutionary game theory	V2V and V2I	Medium	Maximizes nodes payoff	More delay	Advertising, marketing
CGC (Selo et al. 2019)	Coalitional game theory	V2V	Medium	Improves SNR	Stability of the cluster is hard	For reliable and high capacity transmission applications
FAODV (Feyzi and Sattari-Naeini 2015)	Fuzzy logic	V2V	Low	Less end to end delay and more PDR	Does not work for highway scenario	For more stable route applications
FMIFR (Purkait and Tripathi 2019)	Fuzzy logic	V2V and V2I	Medium	Increase link connectivity and throughput	---	Safety to passengers, Internet connectivity
FLDLR (Rana et al. 2019)	Fuzzy logic	V2V	Medium	More efficient in selecting next hop, Lower delivery delay	More data packets cannot be transmitted	Applicable for high speed vehicles

(continued)

TABLE 11.2 (Continued)
Overview of artificial intelligence techniques based routing algorithms

Research Works	AI techniques used	Architecture used	Computational complexity	Advantages	Disadvantages	Applications
GABR (Guoan et al. 2018)	Genetic algorithm	V2V	Complex	Better transmission delay and packet loss rate	Searching speed is less	Applicable for urban scenarios
ACO (Kakkasageri et al. 2019; Salim and Abdelhamid 2014)	Ant colony optimization	V2V	Medium	High quality link between vehicles, high PDR	More delay due to more iterations	QoS based applications
FSO (Sataraddi et al. 2017)	Fish swarm	V2V and V2I	Low	High bandwidth utilization, more PDR	Only speed and distance of vehicles are considered	Applicable for low dense networks
PbQR (Yanglong et al. 2019)	Q-learning	V2V	Medium	More stable and reliable route	Applicable only for low speed vehicles	UDP/CBR based applications
MARS (Wei 2015)	Machine learning	V2I	Medium	High transmission capacity, PDR	Applicable only for high mobile vehicles	---
VANET-QoS-OLSR (Khatib et al. 2015)	Artificial Neural Network	V2V	Medium	Reduces false judgments	More delay	Trust based applications
BDIAR (Sataraddi and Kakkasageri 2017)	BDI agent	V2V	Medium	Increased network lifetime and PDR	Unidirectional movement of vehicles is considered	Safety applications
MAR (Samira et al. 2013)	Multiagent	V2V	Medium	Stable route	More delay	For high traffic networks

- Processing Unit (GPUs). These RSUs perform better computations for artificial intelligence algorithms.
- In software agent based systems, result of fairness in agent's policy is a major challenge in the field of reinforcement learning. Appropriate fairness quantity selection is of prime concern.
- Development of trust building is a major concern for artificial intelligence driven vehicular safety applications.
- Another main area need to be considered is the privacy and security improvement while considering the vehicle density. Even though anomalies and attacks in VANETs need to be discovered by the artificial intelligence algorithms, more knowledge and suitable rules need to be identified to enhance the detection accuracy.
- Artificial intelligence algorithms based VANET require higher computational resources.

This chapter presented comprehensive survey on artificial intelligence techniques based VANET routing. We have presented various artificial intelligence techniques along with the importance of routing in VANETs. Various artificial intelligence techniques based routing protocols for VANETs along with computational models are also discussed. The chapter also addressed the open research issues need to be addressed for deploying artificial intelligence techniques based VANET routing.

REFERENCES

Alsharif, N., and Shen, X. 2017. Citation: iCAR-II: Infrastructure-Based Connectivity Aware Routing in Vehicular Networks. *IEEE Transactions on Vehicular Technology*, 66(5): 4231–44.

Chen, R., Jin, W. L., and Regan, A. 2010. Citation: Broadcasting Safety Information in Vehicular Networks: Issues and Approaches. *IEEE Journal on Network*, 24(1): 20–5.

Chukwu, J., and Agwu, J. N. 2015. Issues and Possibilities in Vehicular Ad-Hoc Networks (VANETs). In *IEEE International Conference on Computing, Control, Networking, Electronics and Embedded Systems Engineering*, pp. 254–9.

Clausen, T., and Jacquet, P. 2003. Optimized Link State Routing Protocol (OLSR). *RFC 3626 (Experimental)*.

Cunha, F., Villas, L., and Boukerche, A. 2016. Citation: Data Communication in VANETs: Protocols, Applications and Challenges. *Elsevier Journal of Ad Hoc Networks*, 44: 90–103.

Daniel, C. 2015. Evolution and Evolutionary Algorithms. *Bio-inspired Networking*, Elsevier.

Dasagoudar, M. K., and Manvi, S. S. 2008. QoS Based Routing in MANETs: ANN Based Approach. In *International Conference on Advance Computing*, pp. 526–9.

Di, W., Jing, C., Yan, L., Jiangchuan, L., and Limin, S. 2012. Citation: Routing Algorithm Based on Multi-Community Evolutionary Game for VANET. *Journal of Networks*, 7(7): 1106–15.

Eberhart, R. C., and Shi, Y. 2007. Computational Intelligence: Concepts to Implementations. Morgan Kaufmann.

El Khatib, A., Mourad, A., and Otrok, H. 2015. A Cooperative Detection Model Based on Artificial Neural Network for VANET QoS-OLSR Protocol. In *IEEE International Conference on Ubiquitous Wireless Broadband*, pp. 1–5.

Feyzi, A., and Sattari-Naeini, V. 2015. Application of Fuzzy Logic for Selecting the Route in AODV Routing Protocol for Vehicular Ad Hoc Networks. In *23rd IEEE Iranian Conference on Electrical Engineering*, pp. 684–7.

Guoan, Z., Min, W., Wei, D., and Xinming, H. 2018. Citation: Genetic Algorithm Based QoS Perception Routing Protocol for VANETs. *Hindawi Journal of Wireless Communications and Mobile Computing*, 1–10.

He, J., Cai, L., Pan, J., and Cheng, P. 2017. Citation: Delay Analysis and Routing for Two-Dimensional VANETs Using Carry-and-Forward Mechanism. *IEEE Transactions on Mobile Computing*, 16(7): 1830–41.

Hridya, P. C., Budyal, V. R., and Kakkasageri, M. S. 2016. Cognitive Agent Based Stable Routing Protocol for Vehicle-to-Vehicle Communication. In *IEEE Annual India Conference (INDICON)*, pp. 1–5.

Janusz, K. and Witold, P. 2015. *Springer Handbook of Computational Intelligence*. Springer.

Johnson, D., Hu, Y., and Maltz, D. 2007. The Dynamic Source Routing Protocol (DSR) for Mobile Ad Hoc Networks for IPv4. *RFC 4728 (Experimental)*.

Kakkasageri, M. S., Manvi, S. S., Hridya, C. P., Vibha, N. S., and Basarkod, P. I. 2012. Citation: A Survey on Cognitive Agent Architecture. *IETE Journal of Education*, 53(1): 21–37.

Kakkasageri, M. S., Sataraddi, M. J., Chanal, P., and Kori, G. 2017. BDI Agent Based Routing Scheme in VANETs. In *IEEE International Conference on Wireless Communications Signal Processing and Networking*, pp. 129–33.

Kakkasageri, M. S., Sataraddi, M. J., Jahagirdar, R., Chanal, P. M. 2019. Performance Analysis of Ant Colony Based Routing Approach for VANETs using Vanet MobiSim and NS2. In *IEEE International Conference on Advanced Computing*, pp. 129–33.

Kapade, N., and Deshmukh, S. 2013. Comparative Study of Game Theoretic Approaches in Distributed System. In *IEEE conference on Computer and Information Technology*, pp. 1–4.

Kenneth, C. 2012. *Introduction to Game Theory:* Oxford University Press.

Ketchpel, S. 1993. Coalition Formation among Autonomous Agents. In *European Workshop Modeling Autonomous Agents in a Multi-Agent World*, pp. 73–88.

Kromer, P., and Musilek, P. 2015. Bio-Inspired Routing Strategies for Wireless Sensor Networks. *Propagation Phenomena in Real World Networks*, pp. 155–81.

Li, F., and Wang, Y. 2007. Routing in Vehicular Ad hoc Networks: A Survey. *IEEE Vehicular Technology*, pp. 12–22.

Lu, R., Lin, X., Zhu, H., and Shen, X. 2010. Citation: Pi: A Practical Incentive Protocol for Delay Tolerant Networks. *IEEE Transactions on Wireless Communications*, 9(4): 1483–93.

Manvi, S. S., and Kakkasageri, M. S. 2008. Citation: Issues in Mobile Ad Hoc Networks for Vehicular Communication. *IETE Technical Review*, 25(2): 59–72.

Mezher, A. M., and Igartua, M. A. (2017) Citation: Multimedia Multimetric Map-Aware Routing Protocol to Send Video-Reporting Messages over VANETs in Smart Cities. *IEEE Transactions on Vehicular Technology*, 66: 10611–25.

Miri, S. T., and Tabatabaei, S. 2020. Citation: Improved Routing Vehicular Ad-Hoc Networks (VANETs) Based on Mobility and Bandwidth Available Criteria Using Fuzzy Logic. *Wireless Personal Communication*, 113: 1263–78.

Morgan, Y. L. 2010. Notes on DSRC and WAVE Standards Suite: Its Architecture, Design, and Characteristics. *IEEE Communications Surveys and Tutorials*, 12(4): 504–18.

Moustafa, H. and Zhang, Y. 2009. *Vehicular Networks: Techniques, Standards and Applications:* Auerbach Publications.

Muhammad, B., Pauline, M., and Chan, L. 2011. A Game Theoretic Coalitional Bidding Scheme for Efficient Routing in Vehicular Ad Hoc Networks. In *IEEE 10th International Conference on Trust, Security and Privacy in Computing and Communications*, pp. 1638–45.

Muhammad, M., Imran, R., and Syed, H. 2017. Citation: A Game Theory Based Trust Model for Vehicular Ad Hoc Networks (VANETs). *Journal of Computer Networks*, 121, 152–72

Nicolas, P., Rafael, F., and Rami, A. 2018. Citation: A Review of Computational Intelligence Techniques in Wireless Sensor and Actuator Networks. *IEEE Journal of Communications Surveys and Tutorials*, 20(4): 2822–54.

Oubbati, O. S., Lakas, A., Lagraa, N., and Yagoubi, M. B. 2015. CRUV: Connectivity Based Traffic Density Aware Routing Using UAVs for VANETs. In *IEEE International Conference on Connected Vehicles and Expo*, pp. 68–73.

Perkins, C., Belding-Royer, E. and Das, S. 2003. Ad Hoc On-Demand Distance Vector (AODV) Routing. *RFC 3561 (Experimental)*.

Purkait, R., and Tripathi, S. 2019. Citation: Fuzzy Logic Based Multi-Criteria Intelligent Forward Routing in VANET. *Wireless Personal Communication*, 111: 1871–97.

Rahul, M., Narinder, S., and Yaduvir, S. 2011. Citation: Genetic Algorithms: Concepts, Design for Optimization of Process Controllers. *Journal of Computer and Information Science*, 4(2).

Rana, K. K., Tripathi, S. and Raw, R. S. 2019. Fuzzy Logic-Based Directional Location Routing in Vehicular Ad Hoc Network. In *National Academy of Sciences, India, Section A: Physical Sciences.*

Sabouni, R., and Hafez, R. M. 2012. Performance of DSRC for V2V Communications in Urban and Highway Environments. In *25th IEEE Canadian Conference on Electrical and Computer Engineering (CCECE), Montreal, QC*, pp. 1–5.

Salim, B., and Abdelhamid, M. 2014. Bio-Inspired Routing Protocols for Vehicular Ad Hoc Networks. *John Wiley & Sons*, pp. 29–50.

Salim, B., and Abdelhamid, M. 2014. Bio-Inspired Routing Protocols for Vehicular Ad Hoc Networks. *John Wiley & Sons*, pp. 79–119.

Samira, H., Walid, C., and Khaled, G. 2013. A Multi-Agent Approach for Routing on Vehicular Ad-Hoc Networks. In *4th International Conference on Ambient Systems, Networks and Technologies*, pp. 578–85.

Sandeep, K., and Medha, S. 2012. Citation: Convergence of Artificial Intelligence, Emotional Intelligence, Neural Network and Evolutionary Computing. *International Journal of Advanced Research in Computer Science and Software Engineering (IJARCSSE)*, 2(3).

Sataraddi, M. J., Kakkasageri, M. S., Kori, G. S., and Patil, R. V. 2017. Intelligent Routing for Hybrid Communication in VANETs. In *7th IEEE International Advance Computing Conference*, pp. 385–90.

Sataraddi, M. J., and Kakkasageri, M. S. 2017. BDI Agent Based Dynamic Routing Scheme for Vehicle-to-Vehicle Communication in VANETs. In *IEEE International Conference on Smart Technology for Smart Nation*, pp. 665–70.

Selo, S., Sahirul, A., and Ronald, A. 2019. Citation: Coalitional Game Theoretical Approach for VANET Clustering to Improve SNR. *Hindawi Journal of Computer Networks and communications*, 1–13.

Shah, S., Shiraz, M., Nasir, K., and Noor, M. 2014. Citation: Unicast Routing Protocols for Urban Vehicular Networks: Review, Taxonomy and Open Research Issues. *Journal of Zhejiang University Science*, 15(7): 489–513.

Sultan, S., Doori, M. M., Bayatti, A. H., and Zedan, H. 2014. Citation: A Comprehensive Survey on Vehicular Ad-Hoc Network. *Elsevier Journal of Network and Computer Applications*, 37: 380–92.

Sungwook, K. 2016. Citation: Timed Bargaining-Based Opportunistic Routing Model for Dynamic Vehicular Ad Hoc Network. *EURASIP Journal on Wireless Communications and* Networking, 14, 1–9.

Sutton, R. S. and Barto, A. G. 2018. *Reinforcement Learning: An Introduction. 2nd edition.* MIT Press.

Tian, D., Zheng, K., Zhou, J., Duan, X., Wang, Y., Sheng, Z., and Ni Q. 2018. Citation: A Microbial Inspired Routing Protocol for VANETs. *IEEE Internet of Things*, 5(4): 2293–2303.

Tingting, C., Liehuang, Z., Fan, W., and Sheng, Z. 2011. Citation: Stimulating Cooperation in Vehicular Ad Hoc Networks: A Coalitional Game Theoretic Approach. *IEEE Transactions on Vehicular Technology*, 60(2): 566–79.

Torrent-Moreno, M., Mittag, J., Santi, P., and Hartenstein, H. 2009. Citation: Vehicle-to-Vehicle Communication: Fair Transmit Power Control for Safety-Critical Information. *IEEE Transactions on Vehicular Technology*, 58(7): 3684–3703.

Vidushi, S., and Anoop, K. 2013. Citation: A Comprehensive Study of Fuzzy Logic. *International Journal of Advanced Research in Computer Science and Software Engineering (IJARCSSE)*, 3(2).

Wadea, M., Mostafa, A., and Hamad, A. 2017. Connecting the Autonomous: A Distributed Game Theory Approach for VANET Connectivity. In *IEEE 86th Vehicular Technology Conference*, pp. 1–5.

Wahid, I., Ikram, A., Ahmad, M., Ali, S., and Ali, A. 2018. Citation: State of the Art Routing Protocols in VANETs: A Review. *Elsevier Procedia Computer Science*, 130: 689–94.

Wang, T., Azhar, H., Wang, B., Sabita, M., and Abita, M. 2018. Citation: Artificial Intelligence for Vehicle-to-Everything: A Survey, *IEEE Access*, 4: 1–22.

Wang, X., Liu, W., Yang, L., Zhang, W., and Peng, C. 2016. A New Content-Centric Routing Protocol for Vehicular Ad Hoc Networks. In *22nd Asia-Pacific Conference on Communications*, pp. 552–8.

Wei, K., Mei-T., and Yu-Hsuan,Y. 2015. Citation: A Machine Learning System for Routing Decision-Making in Urban Vehicular Ad Hoc Networks. *Hindawi International Journal of Distributed Sensor Networks*, 1–13.

Xin-SheY. (2014). Genetic algorithms. *Nature-Inspired Optimization Algorithms, Elsevier*

Yanglong, S., Yiming, L., and Yuliang, T. 2019. Citation: A Reinforcement Learning-Based Routing Protocol in VANETs. *Springer Nature*, 2493–2500.

Yujie, T., Nan, C., and Wen, W. 2019. Citation: Delay-Minimization Routing for Heterogeneous VANETs with Machine Learning Based Mobility Prediction. *IEEE Transactions on Vehicular Technology*, 68(4): 3967–79.

Zoubir, M. 2019. Citation: Reinforcement Learning Based Routing in Networks: Review and Classification of Approaches. *IEEE Access*, 7: 55916–50.

Zrar Ghafoor, K., Abu Bakar, K., and Van Eenennaam, M. 2013. Citation: A Fuzzy Logic Approach to Beaconing for Vehicular Ad Hoc Networks. *Telecommunication Systems*, 52: 139–49.

12 A Comparison of Different Consensus Protocols

The Backbone of the Blockchain Technology

Arvind Panwar and Vishal Bhatnagar

CONTENTS

12.1   INTRODUCTION

Haber and Stornetta introduces blockchain and recent blockchain has been measured as one of the most potential technologies (Grant Thornton 2016). Bitcoin is the first cryptography-based digital currency, and has the power to resolve the old payment method problem, in which one has to trust any third party for payment (Banerjee Mandrita, Lee Junghee 2017; Lee, Azamfar, and Singh 2019). Generally, when any-body wants to make a payment, they need confidence in a third party, who would con-firm the validity of transactions, before pushing payment into action. Inappropriately, this kind of middle party is always suspected regarding whether they might cheat their user or not. This problem derives from recognized centralization, in which the whole thing is governed by the sole organization, causing acceptance to be inadequate (H ES-SAMAALI and Kodama 2018; Risius and Spohrer 2017).

After taking the motivation from this idea, there is a ledger of records in bitcoin and other digital currency or late alternates of blockchain which record every single transaction which has been verified successfully. For example, X sends 50 rupees to Y, Y sends 100 rupees to Z after seing the last ledger, and this is likely to know how many rupees a person has. To make the system trustworthy, with the help of a decentralized method, the ledger is maintained by serval organizations, known as parties or nodes in blockchain (Sousa, Bessani, and Vukolic 2018). A researcher Satoshi has offered a novel design of this kind of ledger, which is known as a block, which stores all successfully validate transactions. The above-mentioned ledger is re-generated and holds several blocks with many transactions, making a chain, which is why we called this kind of ledger a blockchain (Dwivedi, Srivastava, Dhar, and Singh 2019; Janssen, Weerakkody, Ismagilova, Sivarajah, and Irani 2019; Pahl, El Ioini, and Helmer 2018). The very opening block is known as the genesis block in the blockchain, which contains the very first transaction in the blockchain.

Whenever anybody initiates any transaction in blockchain, it can be verified by any node. If it finds it valid, that is, the sender or initiator has sufficient money for the transaction, and the sender also confirmed the transaction by signing her or his digital signature, then this transaction will be added into a block (Biswas, Sharif, Li, Nour, and Wang 2019; Rathee, Sharma, Saini, Kumar, and Iqbal 2020). Now, to mark a transaction valid, a block must be added to the chain that contains this transaction, and all the nodes in the chain must recognize this block. A node always tries to be added to a block containing several transactions in chain by broadcasting it to the rest nodes in chain, which suggests this block should be added in the current chain. However, this form of transaction verification creates a confusion when every node broadcasts their block. To overcome this kind of situation, an agreement should be prepared among all the nodes for which nodes are allowed to add their blocks, and which block added to the chain. This kind of agreement is known as a consensus algo-rithm (Luu et al. 2016). Until now many consensus algorithms have been proposed by researchers.

To get motivation from previous research, the first proof based variant consensus algorithm has been proposed and is known as proof of work (PoW) consensus (Bai and Sarkis 2020). On the basic of drawbacks of PoW algorithm a new algorithm is proposed known as PoS (proof of stake) which pays the stake of every node, and a block is added to the chain which depends upon the lucky factor (Zheng, Xie, Dai, Chen, and Wang 2017). Until now, both PoS and PoW based algorithm have been highly popular in the market to be used in application and research, but there is another proposed algorithm also such as proof of space, proof of luck, elapsed time.

12.1.1 Block Diagram of Blockchain

A blockchain can be defined as a data structure which holds some transaction records and ensuring transparency, security, and decentralization. In other words, we can also consider blockchain as records or a chain stored in the forms of some kind of block which are not controlled by any single administration. Figure 12.1 shows the basic block diagram of blockchain. This figure shows how a transaction is complete in a blockchain.

12.1.2 Consensus Algorithms

As we all know, blockchain is a peer-to-peer decentralized system which doesn't have any kind of central authority (Swan 2018; Zheng, Xie, Dai, Chen, and Wang 2018). While we create a system like this, which is free from corruption of a single source, it still creates substantial problems and raises questions such as: how are any choices made and how does something get done? Now consider a normal centralized system, where all choices are made by a leader. These kinds of choices are not possible in a decentralized system because there is no such leader. To overcome this kind of problem or situation in blockchain we need to come with a consensus using consensus algorithms (Kumar and Mallick 2018; Pan, Pan, Song, Ai, and Ming 2020). So, consensus algorithm is known as a decision-making technique in blockchain for a cluster, where people of every cluster support and construct the pronouncement which works finest for the others. It is a kind of resolution, where every specific person must back the widely held pronouncement, whether anybody likes it or does not like it. Every blockchain consensus algorithm has some objectives which are given below:

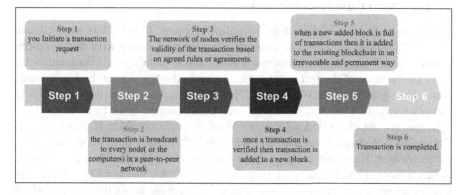

FIGURE 12.1 Working of blockchain.

An agreement: Every consensus algorithm collects all the agreements from all the participants of the group and makes a final agreement on the basic of majority decision which will be followed by all members of the group.

- *Collaboration:* every member of a group wants betterment of all members in group and has a single aim toward better agreement for the group which covers everybody's interest.
- *Co-operation:* every member of cluster will work as a crew for good results and place their individual interest away.
- *Equal rights*: Each person in a group has equal rights to vote or have identical value for voting. This means each individual is equally important in a cluster to make an agreement.
- *Participation:* Participation of every person is compulsory in the voting process; nobody cannot be left for voting or cannot stay away from the voting process.
- *Activity:* every person's activity matters in the group. Every person is similarly active in group. There is nobody with higher charge or accountability in the group.

12.1.3 THE PROBLEM WITH BYZANTINE FAULT TOLERANCE

Now a question raised is why distributed ledger technology or blockchain needs a consensus mechanism? And the answer is the byzantine problem which is very common in distributed computing (Di Vaio and Varriale 2019). Let us suppose there is a crowd of byzantine generals and they want to launch an attack on a state. At the time, they are facing different problems:

- Armies and generals are very far away from central power or authority, and it is impossible to coordinate the attack.
- The state has a very huge army and generals can only win when they all attack at the same time.

12.1.4 NEED CONSENSUS ALGORITHMS

In the last section we explained the byzantine problem. To solve this we need consensus algorithms. Let's take an example to understand clearly what this byzantine problem is. In order to mark a successful attack and to make a good coordination, the army on the right of the fort sends a messenger to the left side of the fort's armies with a message "Attack on Wednesday." Now let us suppose, the armies on left side of fort is not ready for the attack and send back the messenger to the right side with the message "No, attack on Friday." They will face now face a problem (Lai and Lee Kuo Chuen 2017; Rodrigo, Senaratne, and Weinand 2020). Many things could happen to the messenger. He or she could get compromised, be replaced with another person, or be captured or and killed by the state (Kshetri 2017; Li, Wu, and Chen 2018). This kind of activity can mislead the armies and they can get the wrong information, which means the attack will not be done at once or at the same time or in coordination and the armies will be defeated or lose the war against the state (Zhou, Fu, Yu, Su, and Kuang 2018). This can happen in blockchain. Blockchain is a very huge network, so how can anybody trust anyone else? If Alice sends 40 bitcoin to Bob from a wallet, how can Bob be sure that anybody in the network will not tamper with the transaction and change 40 bitcoins to 4 bitcoins? For this we need some consensus algorithms.

12.2 BLOCKCHAIN TECHNOLOGY

Let's dive into blockchain technology to understand the whole network.

- Blockchain is a new method to organize database.
- Blockchain can store anything which change network.
- Block is used to arrange the data in Blockchain.
- All data transaction is link with next and previous transaction with hash value to create a chain.

12.2.1 CONSENSUS MECHANISM: THE BACKBONE OF THE NETWORK

Consensus mechanism or protocols are the backbone or you may say the heart of the blockchain network. Blockchain is a decentralized network but nobody can see any kind of method in the blockchain (Babich and Hilary 2020; Khan and Salah 2018). This has happened because it does not offer a decentralized environment; to offer decentralization a blockchain network needs consensus algorithms. Blockchain technology only offers to create a diverse structured database, and it would not carry any kind of decentralization process. That's why blockchain is known as only the skeleton of a decentralized system and consensus algorithm is known as the backbone or heart of the blockchain system (Reyna, Martín, Chen, Soler, and Díaz 2018). The technique is really very simple: to provide the decentralization by using a consensus model. Blockchain consensus algorithms are only the agreement to perform any task or verification of any kind of transaction (Milutinovic, He, Wu, and Kanwal 2016). There are many different kinds of consensus available, some of which will be explained in the next section.

12.2.2 DIFFERENT TYPES OF CONSENSUS PROTOCOLS

Figure 12.2 shows different type of blockchain consensus algorithms.

12.3 WORKING OF DIFFERENT TYPES OF CONSENSUS PROTOCOLS

Consensus protocol is the backbone of blockchain technology to provide transparency and agreement in the participating node. The technique is very simple and straightforward. These consensus mechanisms or protocols are the best technique to reach a treaty. However, decentralized system can't work without a consensus mechanism

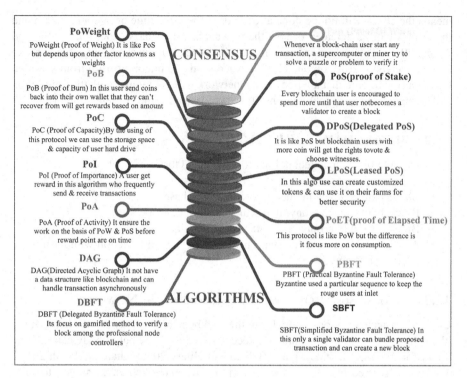

FIGURE 12.2 Different types of consensus algorithm.

(Filimonau and Naumova 2020; Muzammal, Qu, and Nasrulin 2019). It won't matter whether or not the participating nodes in a distributed network trust one another or not. They'll need to waiver bound principles and reach joint agreement. To achieve this, we have to go study all the consensus protocol.

12.3.1 PoW (PROOF-OF-WORK)

This is the primary consensus mechanism used in distributed ledger technology for consensus mechanism. The very first application of blockchain known as bitcoin is based upon proof of work consensus mechanism. Bitcoin uses a PoW algorithm to confirm transaction in decentralized environment and generate new blocks according to the consensus in the blockchain network.

The distributed ledger system gathers all the data belonging to blocks from the complete blockchain. However, some participating nodes have to take superior care of all the transactions performed by blocks (Peters and Panayi 2016). The participating node, which takes special care of the transaction, is known as the Miner node and the task they perform is known as mining. The fundamental principle of PoW algorithm is to resolve a very complex cryptographical puzzle or you can say a mathematical problem and find a solution. Figure 12.3 shows the working of PoW.

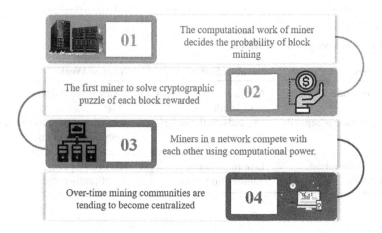

FIGURE 12.3   Proof of work in bitcoin.

12.3.2   PoS (Proof-of-Stake)

In the last section we discussed PoW which is very effective and shows good potential in bitcoin cryptocurrency application. But there are some major issues with bitcoin, including energy consumption: 51 percent attack and mining process depend upon the computation power of a computer (Meng, Tischhauser, Wang, Wang, and Han 2018). Proof of stake is proposed to tackle these problems. PoS is one of the best consensus mechanisms for distributed technology.

The basic concept of PoS is that any participating node can validate or mine new blocks depend upon the stake they hold in wallet or we can say how many coins a node has. In this case, the more coins you have the more chance to mine a block and get a reward.

In PoS consensus algorithm, miner nodes are chosen in the earlier phase. The procedure to select miner node is completely non-consecutive. In PoS, each selected miner does not join stake processing because stake processing is dependent upon a certain number of coins a particular node has in the wallet. If a miner has less coin in wallet, he can join as a node in the distributed network but if he wants to be part of the stake processing he needs to deposit some more coins to qualify for the certain number of coins criteria. After depositing coins, there will be a voting process to select validators from node. The complete working is shows in Figure 12.4 and Table 12.1 shows the difference between PoW vs. PoS.

12.3.3   DPoS (Delegated Proof-of-Stake)

DPoS protocol is the modified version of classic PoS consensus protocol. DPoS is pretty strong and further adds some more flexibility in the complete distributed system in comparison with classical PoS. If your organization demands efficient, decentralized consensus mechanism and fast processing, then you must choose DPoS

TABLE 12.1
Proof of work vs proof of stake

Parameters	Proof of Stake	Proof of Work
Validating node	Nodes are called forgers or validators.	Nodes are called miners.
Block generation capacity	Depends on the stake	Depends on computational power.
Coins	No new coins	Produces new coins.
Reward	Receive transaction fees.	Receive block rewards.
energy consumption.	Low to moderate	Massive
51% attacks	almost impossible.	Significantly prone

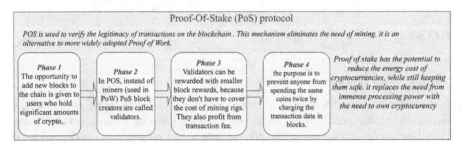

FIGURE 12.4   Proof of stake.

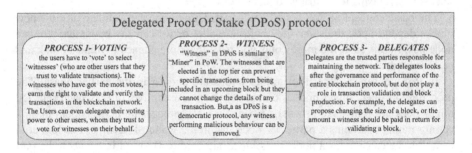

FIGURE 12.5   Working of delegated proof-of-stake.

for the best result. In this mechanism, any participating node can become delegate node. And in DPoS if a node working like validator or miner is known as delegate node, DPoS can complete a transaction in a single second as compared to PoW which takes ten minutes to complete transaction. The complete working and function of DPoS can be divided into three processes as shown in Figure 12.5.

12.3.4   LPoS (Leased Proof-of-Stake)

This consensus protocol is designed by waves organization. Waves modify the standard PoS consensus algorithm to develop LPoS. Waves are a new blockchain

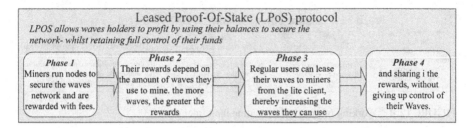

FIGURE 12.6   Working of leased proof-of-stake.

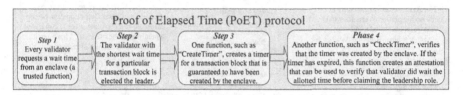

FIGURE 12.7   Working of proof of elapsed time.

technology platform which uses LPoS to attain the distributed consensus, transparency, and security in network. In LPoS, participating nodes use leased token to create a block and collect reward. To do this task a particular node must have at least 1000 waves. A unique feature of LPoS is token leasing to other nodes by token holder to get a reward. Figure 12.6 shows the working of LPoS.

12.3.5   POET (PROOF OF ELAPSED TIME)

This is one of the best consensus mechanisms for distributed network. PoET is specially planned for private blockchain network or permissioned blockchain where every participating node needs permission to use the network. Permission granted to nodes depends upon voting principles and mining rights. PoET consensus mechanism uses special time base method to make sure smooth running of consensuses and provide transparency in the network. Every participating node needs to delay for an arbitrary quantity of time in the network. And after any node finished its waiting time, then the node can create a block. Figure 12.7 shows the working of PoET.

12.3.6   PBFT (PRACTICAL BYZANTINE FAULT TOLERANCE)

This attempts to deliver a sensible byzantine state mechanism reproduction which will work under a condition in which a malicious participating node is operating in the network. There are two different types of node working in PBFT enabled distributed network: (1) leader node or primary node and (2) backup node or secondary node. Any eligible secondary node can become primary node by the conversion from secondary to primary node whenever the primary node fails to fulfil its duty. The working of PBFT is shown in Figure 12.8.

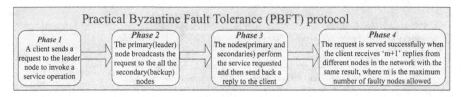

FIGURE 12.8 Working of practical byzantine fault tolerance.

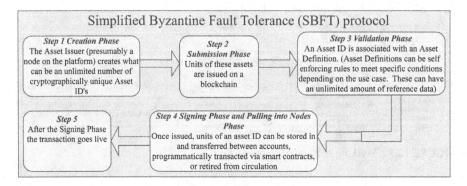

FIGURE 12.9 Working of simplified byzantine fault tolerance.

12.3.7 SBFT (SIMPLIFIED BYZANTINE FAULT TOLERANCE)

In this protocol, the distributed network works differently. Initially, a block generator accumulates all the transactions in a single block at a time and makes a batch of all transactions and then authenticates all transaction collectively in a new kind of block. The block generator sets some protocols, which all participating node have to follow to authenticate any transaction. Figure 12.9 shows the working of SBFT.

12.3.8 DBFT (DELEGATED BYZANTINE FAULT TOLERANCE)

This consensus mechanism completely depends upon PDFT consensus mechanism, which is best suited to solving consensus effectively in distributed ledger network. But whenever the number of participating nodes in distributed network increases, the performance drops quickly, because the time complexity is $O(n^2)$. This is the only reason NEO developed the new consensus algorithm DBFT with the combination of PBFT and DPoS characteristics. Figure 12,10 shows the working of DBFT.

12.3.9 DAG (DIRECTED ACYCLIC GRAPHS)

This is a topological ordering-based data structure. Many cryptocurrency experts accept that Blockchain 1.0 is bitcoin and Blockchain 2.0 is Ethereum. However, many new players come in the cryptocurrency market with new technology. Some researchers says DAG is Blockchain 3.0 and NXT is using DAG which changes the market and showing the great potential to lead in this area. IoT Chain, Hashgraph, and IoTA is also using DAG for distributed ledger. In this kind of blockchain consensus

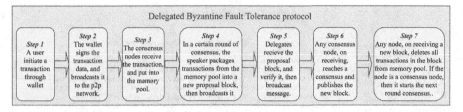

FIGURE 12.10   Working of delegated byzantine fault tolerance.

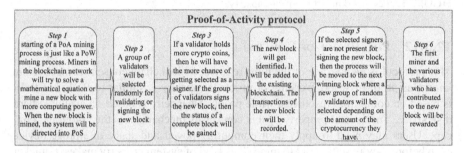

FIGURE 12.11   Working of proof-of-activity.

mechanism, each participating node itself contributes to becoming miners. Now, once miners are exterminated and transactions are validated by users themselves, the related fee decreases to zero. It becomes very easy to authenticate transactions amid any two participating nodes, which makes the full method faster, lightweight, and secure.

12.3.10   PoA (PROOF-OF-ACTIVITY)

Proof-of-Activity is fundamentally a fusion tactic developed by the merging the concepts of PoS and PoW consensus algorithm of Blockchain. In the case of Proof-of-Activity consensus protocol, miner's compete to resolve a cryptographical mystery at the earliest possible time using superior hardware computing resources, similar to PoW. However, the blocks they are available across have only information regarding the individuality of the block front-runner and reward transaction. This is the point, wherever the PoS comes in the picture. Figure 12.11 shows the working of Proof-of-Activity.

12.3.11   PoI (PROOF-OF-IMPORTANCE)

These consensus algorithms are developed by NEM as a new consensus protocol for their cryptocurrency. Proof-of-importance uses the development concept of PoS. In addition, NEM introduced the idea of harvesting or vesting in proof-of-importance protocol. Proof-of-importance uses the sequence of operation that determines which nodes are entitled to add new block in the blockchain network and this process is known as harvesting. To be qualified for harvesting, you need to possess a minimum of 10,000 XEM on your account.

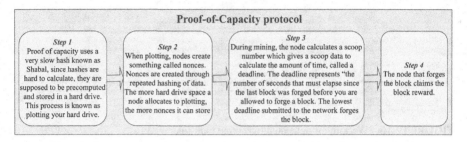

FIGURE 12.12   Working of proof-of-capacity.

12.3.12   PoC (PROOF-OF-CAPACITY)

This protocol is an upgrade of the most renowned blockchain consensus mechanism known as proof-of-work. Working of proof-of-capacity is similar to PoW, but PoW is based on computing resources, while proof-of-capacity is based on available hard disk space. The crucial property of this consensus mechanism is the "plotting" concept. Users may need to apply their computing power and space of storage device even before they're setting out to mine. This kind of system property makes it quicker than POW. Figure 12.12 shows the working of proof-of-capacity protocol.

12.3.13   PoB (PROOF-OF-BURN)

The sequences in this consensus are quite spectacular. To protect the POW cryptocurrency, some of the coins are going to be burnt! The method occurs because the miners send many coins to an associated "eater" address. And this eater node cannot pay these coins for any use. An archive retains track of the burnt coins creating them honestly unspendable. The node that burnt these coins can also get a bequest. Yes, the burning of coins could be a loss. However, the harm is momentary because the method can protect the coins by the end of the day from cyberattacks and hackers. Additionally, the burning method will increase the stakes of the choice coins. Such a state of affairs will increase the prospect of a user to mine following block also as it increases their rewards in future. So, burning might be used as a mining honor. Figure 12.13 shows the working of the PoB protocol.

12.3.14   PoWEIGHT (PROOF-OF-WEIGHT)

PoWeight (proof-of-weight) is a consensus algorithm for blockchain, which provides participants with a "weight" based mechanism on how much weight they are holding at a time. The PoWeight consensus algorithm is based mostly on the primary proof-of-weight consensus mechanism employed in the Algorand cryptocurrency, that was programmed and developed by researchers at the MIT Computer Science Research laboratory. This may be a huge upgrade of the PoS consensus algorithm. In PoS, the more coins the user has, the higher their likelihood is to find extra! This kind of concept makes the structure a touch unfair. The PoWeight tries to unravel this unfairness property of the PoS. Blockchain applications like Filecoin, Chia, and Algorand use the

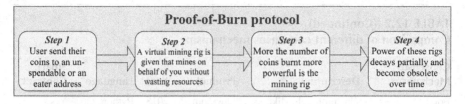

Proof-of-Burn protocol

Step 1	*Step 2*	*Step 3*	*Step 4*
User send their coins to an unspendable or an eater address	A virtual mining rig is given that mines on behalf of you without wasting resources	More the number of coins burnt more powerful is the mining rig	Power of these rigs decays partially and become obsolete over time

FIGURE 12.13 Working of proof-of-burn.

PoWeight. The PoWeight contemplates a different aspect to holding additional coins as in PoS. These aspects are known as the "weighted aspects." The basic benefits of this technique embody customization and measurability. However, incentivizing can be an enormous challenge for this algorithmic rule.

12.4 COMPARISON BETWEEN CONSENSUS MECHANISMS

Table 12.2 shows a comparison between the different consensus mechanisms.

TABLE 12.2
Comparison of different consensus mechanism

Consensus Mechanism	Description	Product	Year	Language	Smart Contract
PoW (proof of work)	Whenever a blockchain user starts any transaction, a supercomputer or a miner tries to crack a mystery or problem to authenticate it	Bitcoin	2009	C++	No
PoS (proof of stake)	Every blockchain worker is invigorated to devote extra until that user does not convert a validator to generate a block	NXT	2013	Java	Yes
DPoS (delegated PoS)	Like PoS but blockchain handlers with extra coin will get the rights to vote and choose witnesses	Lisk	2016	JavaScript	No
LPoS (leased PoS)	In this algo user can create customized tokens and use it on their farms for better security	Waves	2016	Scala	Yes

(continued)

TABLE 12.2 (Continued)
Comparison of different consensus mechanism

Consensus Mechanism	Description	Product	Year	Language	Smart Contract
PoET (proof of Elapsed Time)	This protocol is like PoW but the difference is it focuses more on consumption	Sawtooth	2018	Python, Go, Java, and Rust	Yes
PBFT (practical byzantine fault tolerance)	Byzantine uses a specific arrangement to retain the users at inlet	Hyperledger Fabric	2015	Python, Java Script, Go, C++, Java,	Yes
SBFT (simplified byzantine fault tolerance)	In this only a single validator can bundle proposed transaction and can create a new block	Chain	2014	Java, Ruby	No
DBFT (delegated byzantine fault tolerance)	Its focus on gamified method to verify a block among the professional node controllers	NEO	2016	Python, Java, C, Go, Kotlin,	Yes
DAG (directed acyclic graph)	It does not have a data structure like blockchain and can handle transaction asynchronously	IoTA	2015	Java Go, and C++	Progress
PoA (proof of activity)	It ensures the work on the basis of PoW and PoS before reward points are on time	Decred	2016	Go	Yes
PoI (Proof of Importance)	A user get reward in this algorithm who frequently send and receive transactions	NEM	2015	Java, C++	Yes
PoC (proof of capacity)	By using this protocol we can use the storage space of user hard drive	Burstcoin	2014	Java	Yes
PoB (proof of burn)	In this user send coins back into their own wallet.	Slimcoin	2014	Python, JavaScript	No
PoWeight (proof of weight)	Like PoS but depends upon other factor known as weights	Filecoin	2017	SNARK/ STARK	Yes

12.5   CONCLUSION

Blockchain is a peer-to-peer decentralized system which doesn't have any kind of central authority. While we have created a system like this, which is free of corruption from a single source, it still creates some big problems and raises questions such as: how are any choices made? To overcome this kind of problem or situation in blockchain we need to establish a consensus using consensus algorithms. So, consensus algorithm is known as a decision-making technique in blockchain for a cluster, where people of every cluster support and construct the pronouncement which works finest for others. It is a kind of resolution, where every individual must back the widely held pronouncement, whether anybody likes it or not. Consensus algorithms or protocols are the backbone or you may say heart of blockchain network.

In this chapter we have discussed different type of consensus algorithm. With the help of consensus mechanism, we are able to make a versatile blockchain network. There is not a single algorithm for consensus mechanism in blockchain which can be called perfect. There is always a need for changes to make the system better. There are many researchers, who work on the consensus to make the distributed system more flexible, make transaction faster, and make the system secure. In the near future, the world will see yet more consensus mechanisms for blockchain.

REFERENCES

Babich, V., and Hilary, G. 2020. Distributed Ledgers and Operations: What Operations Management Researchers Should Know about Blockchain Technology. *Manufacturing and Service Operations Management*, 22(2): 223–40. https://doi.org/10.1287/MSOM.2018.0752.

Bai, C., and Sarkis, J. 2020. A Supply Chain Transparency and Sustainability Technology Appraisal Model for Blockchain Technology. *International Journal of Production Research*, 58(7), 2142–62. https://doi.org/10.1080/00207543.2019.1708989.

Banerjee, M., Lee, J., Kwang, K., and R. Choo (2017). A Blockchain Future to Internet of Things Security.pdf. *Digital Communications and Networks*.

Biswas, S., Sharif, K., Li, F., Nour, B., and Wang, Y. 2019. A Scalable Blockchain Framework for Secure Transactions in IoT. *IEEE Internet of Things Journal*, 6(3): 4650–9. https://doi.org/10.1109/JIoT.2018.2874095.

Di Vaio, A., and Varriale, L. 2019. Blockchain Technology in Supply Chain Management for Sustainable Performance: Evidence from the Airport Industry. *Elsevier*. https://doi.org/10.1016/j.ijinfomgt.2019.09.010.

Dwivedi, A. D., Srivastava, G., Dhar, S., and Singh, R. 2019. A Decentralized Privacy-Preserving Healthcare Blockchain for IoT. *Sensors (Switzerland)*, 19(2): 1–17. https://doi.org/10.3390/s19020326.

Es-Samaali, A. O. J. L., Va, N., and Kodama, R. N. T. T. S. 2018. A Blockchain-Based Access Control for Big Data. *International Journal of Computer Networks and Communications Security*, 6(7): 137–47.

Filimonau, V., and Naumova, E. 2020. The Blockchain Technology and the Scope of Its Application in Hospitality Operations. *International Journal of Hospitality Management*, 87. https://doi.org/10.1016/j.ijhm.2019.102383.

Grant Thornton. 2016. A Beginner's Guide to Blockchain. *Grant Thornton*, 1: 1–9. https://doi.org/10.1007/978-0-387-93837-0.

Janssen, M., Weerakkody, V., Ismagilova, E., Sivarajah, U., and Irani, Z. 2019. A Framework for Analysing Blockchain Technology Adoption: Integrating Institutional, Market and Technical Factors. *Elsevier.* https://doi.org/10.1016/j.ijinfomgt.2019.08.012.

Khan, M. A., and Salah, K. 2018. IoT security: Review, Blockchain Solutions, and Open Challenges. *Future Generation Computer Systems*, 82: 395–411. https://doi.org/10.1016/j.future.2017.11.022.

Kshetri, N. 2017. Blockchain's Roles in Strengthening Cybersecurity and Protecting Privacy. *Telecommunications Policy*, 41(10): 1027–38. https://doi.org/10.1016/j.telpol.2017.09.003.

Kumar, N. M., and Mallick, P. K. 2018. Blockchain Technology for Security Issues and Challenges in IoT. *Procedia Computer Science*, 132: 1815–23. https://doi.org/10.1016/j.procs.2018.05.140.

Lai, R., and Lee Kuo Chuen, D. 2017. Blockchain-From Public to Private. In *Handbook of Blockchain, Digital Finance, and Inclusion* (1st ed., Vol. 2). https://doi.org/10.1016/B978-0-12-812282-2.00007-3.

Lee, J., Azamfar, M., and Singh, J. 2019. A Blockchain Enabled Cyber-Physical System Architecture for Industry 4.0 Manufacturing Systems. *Manufacturing Letters*, 20: 34–9. https://doi.org/10.1016/j.mfglet.2019.05.003.

Li, J., Wu, J., and Chen, L. 2018. Block-Secure: Blockchain Based Scheme for Secure P2P Cloud Storage. *Information Sciences*, 465: 219–31. https://doi.org/10.1016/j.ins.2018.06.071.

Luu, L., Narayanan, V., Zheng, C., Baweja, K., Gilbert, S., and Saxena, P. 2016. A Secure Sharding Protocol for Open Blockchains. *Proceedings of the ACM Conference on Computer and Communications Security*, 24–28-Octo, 17–30. https://doi.org/10.1145/2976749.2978389.

Meng, W., Tischhauser, E. W., Wang, Q., Wang, Y., and Han, J. 2018. When Intrusion Detection Meets Blockchain Technology: A Review. *IEEE Access*, 6: 10179–10188. https://doi.org/10.1109/ACCESS.2018.2799854.

Milutinovic, M., He, W., Wu, H., and Kanwal, M. (2016). Proof of Luck: An Efficient Blockchain Consensus Protocol. *SysTEX 2016 – 1st Workshop on System Software for Trusted Execution, Colocated with ACM/IFIP/USENIX Middleware 2016*, 2–7. https://doi.org/10.1145/3007788.3007790.

Muzammal, M., Qu, Q., and Nasrulin, B. 2019. Renovating Blockchain with Distributed Databases: An Open Source System. *Future Generation Computer Systems*, 90: 105–17. https://doi.org/10.1016/j.future.2018.07.042.

Pahl, C., El Ioini, N., and Helmer, S. 2018. A Decision Framework for Blockchain Platforms for IoT and Edge Computing. *IoTBDS 2018 – Proceedings of the 3rd International Conference on Internet of Things, Big Data and Security*, 2018-March (March): 105–13. https://doi.org/10.5220/0006688601050113.

Pan, X., Pan, X., Song, M., Ai, B., and Ming, Y. 2020. Blockchain technology and enterprise operational capabilities: An empirical test. *International Journal of Information Management*, 52. https://doi.org/10.1016/j.ijinfomgt.2019.05.002.

Peters, G. W., and Panayi, E. 2016. Understanding Modern Banking Ledgers through Blockchain Technologies: Future of Transaction Processing and Smart Contracts on the Internet of Money. *New Economic Windows,* 239–78. https://doi.org/10.1007/978-3-319-42448-4_13.

Rathee, G., Sharma, A., Saini, H., Kumar, R., and Iqbal, R. 2020. A Hybrid Framework for Multimedia Data Processing in IoT-Healthcare Using Blockchain Technology. *Multimedia Tools and Applications*, 79(15–16): 9711–33. https://doi.org/10.1007/s11042-019-07835-3.

Reyna, A., Martín, C., Chen, J., Soler, E., and Díaz, M. 2018. On Blockchain and Its Integration with IoT. Challenges and Opportunities. *Future Generation Computer Systems*, 88(2018): 173–90. https://doi.org/10.1016/j.future.2018.05.046.

Risius, M., and Spohrer, K. 2017. A Blockchain Research Framework. *Business and Information Systems Engineering*, 59(6): 385–409. https://doi.org/10.1007/s12599-017-0506-0.

Rodrigo, M. N. N., Senaratne, S., and Weinand, R. 2020. Blockchain Technology: Is It Hype or Real in the Construction Industry? *Journal of Industrial Information Integration*. https://doi.org/10.1016/j.jii.2020.100125.

Sousa, J., Bessani, A., and Vukolic, M. 2018. A Byzantine Fault-Tolerant Ordering Service for the Hyperledger Fabric Blockchain Platform. *Proceedings – 48th Annual IEEE/IFIP International Conference on Dependable Systems and Networks, DSN 2018* (Section 4), 51–8. https://doi.org/10.1109/DSN.2018.00018.

Swan, M. 2018. Blockchain for Business: Next-Generation Enterprise Artificial Intelligence Systems. In *Advances in Computers* (1st ed., Vol. 111). https://doi.org/10.1016/bs.adcom.2018.03.013.

Zheng, Z., Xie, S., Dai, H., Chen, X., and Wang, H. 2017. An Overview of Blockchain Technology: Architecture, Consensus, and Future Trends. *Proceedings – 2017 IEEE 6th International Congress on Big Data, BigData Congress 2017* (June), 557–64. https://doi.org/10.1109/BigDataCongress.2017.85.

Zheng, Z., Xie, S., Dai, H. N., Chen, X., and Wang, H. 2018. Blockchain Challenges and Opportunities: A Survey. *International Journal of Web and Grid Services*, 14(4): 352–75. https://doi.org/10.1504/IJWGS.2018.095647.

Zhou, L., Fu, A., Yu, S., Su, M., and Kuang, B. 2018. Data Integrity Verification of the Outsourced Big Data in the Cloud Environment: A Survey. *Journal of Network and Computer Applications*, 122: 1–15. https://doi.org/10.1016/j.jnca.2018.08.003.

13 Blockchain in AI
Review of Decentralized Smart System

Vikas Goel, Narendra Kumar, Amit Kumar Gupta, and Sachin Kumar

CONTENTS

13.1   INTRODUCTION

13.1.1   ARTIFICIAL INTELLIGENCE

AI is a sub-field of computer science that aims to build computer systems that can perform tasks that often require human ingenuity. For years, the challenging goal of AI has been to develop computer programs that are equal to or beyond human intelligence. AI-based devices are intended to detect their environment and take steps to increase their level of success. AI research uses techniques from a wide range of fields, such as language, economics, and psychology. These techniques are used in applications, such as control systems, natural language processing, facial recognition, speech recognition, business analytics, pattern matching, and data mining.

Blockchain node sharing in device intelligence is used for data integration and recording. A blockchain-based fog node is used to process data on edge fog intelligence. In cloud intelligence, big data is analyzed. The artificial invention provides brilliant dynamics to human space equipment. Gadgets are installed to complete assignments naturally; otherwise, a human mind is needed.

In AI, making powerful intelligent gadgets is based on three levels, great ingenuity, standard AI, and limited AI. Spying machines are used to complete the tasks of people in different areas, for example, medical science, customized gadgets, robotic driving, and car investing [1]; the wisdom of making a computer science branch, responsible for building smart things; who can think intelligently as human beings, can solve problems, and change on their own. Accept raw data as environmental information, make decisions, and perform this function through actors [2].

Starting late many researchers have encountered many difficulties in IoT, for example, big data testing, security, and reliability, gridlock, and information capacity. To address these issues, IoT AI methods have developed, for example, machine learning, in-depth learning created. With in-depth learning use, the energy-saving problem is defined in IoT based on a good structure where critical limitations, for example, data integration, data limitations are severe and are associated with energy management constructions to assist engineers with a critical choice of energy management settings [3].

Machine learning has a vast area of applications such as to identify examples, inconsistencies, and performance expectations based on large amounts of data generated by IoT applications, for example, health services and weather forecasting. Machine-based machine learning authentication provides IoT security solutions to protect data privacy in accordance with machine learning strategies. The machine learning process includes supervised, supervised, and enhanced learning, accessibility, secure uploads, and malware detection strategies. [4,5].

The scientists recognize designs and give more educated choices to another biological system by applying the examination abilities in AI. Figure 13.1 shows the essential methods utilized in AI.

13.1.2   BLOCKCHAIN

Blockchain is a chain of blocks that are connected by the hash value of the previous block where each block stores the following information:

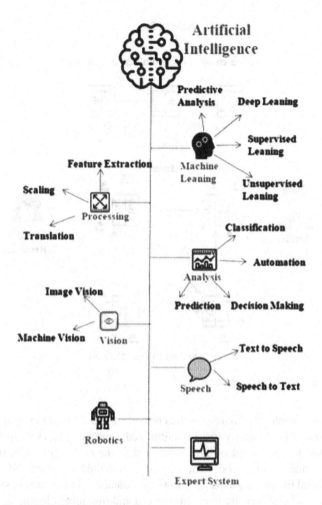

FIGURE 13.1   Artificial intelligence techniques.

- transaction's details (bitcoin, ethereum),
- the calculated hash value of the current block and the previous block, and
- the timestamp of the transaction.

Blockchain technology often referred to as the next big digital revolution after the advent of the Internet, is a common purpose application of existing technology, called "Digital Ledger Technology" (DLT). This technology allows everyone who is part of the blockchain network to verify transaction details (for example, who paid for whom at the time) from a standard non-database database) where all recorded activity is "distributed" to all system participants, who verify the authenticity of the transaction. This eliminates the need for a consultant (such as a bank) who must act as a guaranteed authority. Everything that is done in blockchain technology is signed with a hash value secretly. The transaction is also guaranteed by all the miners; as shown in Figure 13.2.

Blockchain

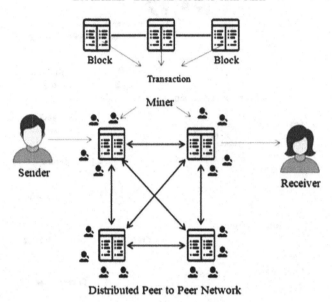

FIGURE 13.2 Blockchain concepts.

Blockchain enables the sharing of data records in the distribution, transfer, protection, and trust of the property [6]. Decentralized power is a blockchain method and has been used to place a lot of data connected to the current block in the previous block with a smart contract code. The majority, LitecoinDB, MoneroDB, SiacoinDB, IPFS, BigchainDB, etc. are used for a low-level database in the current situation [6].

The Point to Point website is medium-sized and distributed by the Interplanetary File System (IPFS). This file system is connected and used to transfer standard files [7,8]. In blockchain technology, IPFS is an important endpoint used for high-performance IoT systems [9]. Atzori et al. [1] has identified three classes of IoT: the web is organized, the sensor is accessible, and the information is organized. Web set up IoT refers to the integration of web-related gadgets, forming a huge amount of data. The sensor available to IoT is the use of a sensor-related frill, for example, RFID. Data designed for IoT refers to the integration of trained data and is used for IoT programs. IoT is a new development with a combination of various gadgets that have a unique location and communicate with each other using the web all this time. For IoT applications, due to a large amount of data, various challenges pose security, privacy, and error tolerance challenges. To measure this difficulty, many experts suggest the creation of a new IoT blockchain.

Blockchain-related approaches offer a broader and more comprehensive structure for the security, vulnerability, and protection as well as adaptation to internal failures; including significant power consumption, computer deployment, and minimal reversal; security and IoT protection are provided by a new blockchain for a variety of applications, for example, smart city, medical care, and agribusiness. The blockchain concept provides a shared network of verification, power against attacks. The installation of blockchain and IoT has a holistic approach and is used for a comprehensive data board and optimization environment with multiple features.

13.1.3 BLOCKCHAIN AND AI

Blockchain is a new disruptive technology that enables digital currencies such as bitcoin and refers to the continuous continuation of records or records that are connected and verified through cryptography [10,11]. More recently, it has been outsourced to the protection of the Internet, IoT, advanced records, and other new data. From now on, a lot of testing and speculation is being put into the head of the new blockchain, due to the great potential and endless performance it offers. Besides, the new blockchain is becoming the most important leap forward since the advent of the Internet [10,11] – and it needs to become an integral part of many organizations. The new blockchain capability to ensure data accuracy makes it useful for a few AI applications, both taking care of AI framework data and recording results from them. The invention of AI or machine learning has been with us for a very long time. It is stylish, as can be seen in how bots virally take control over the webspace.

Today, AI is on the rise in ever-increasing use and is inseparably incorporated into all aspects of daily life, taking importance in almost every area of the economy, from budget management and banking, medical care, transportation, care among others. Moreover, with the extensive and unparalleled installation of AI applications, the authors are at the beginning of a computer world controlled by a two-machine interface or machine economy. Furthermore, it is undeniable that AI and blockchain are two important developments that promote developmental movement and produce extreme movement in each area of the business. In addition to this, everything new has its natural level with various features such as business preparation but the combined use of the two can have the option of reviving the whole (human) world view without preparation. Part of the organization working on implementing this development includes Cognitive Scale, an AI startup supported by IBM, Intel, Microsoft, and USAA, among others looking to use blockchain technology to securely store the results of an AI system that works for management consistency in the business environment. An industry restricted by a ton of management, in this way with the option to keep AI-managed options secure should help market members to stabilize beyond sensitive critical needs.

IBM is also now playing the marriage of the two that offer the blockchain – thanks to the open-source Hyperledger Fabric – and its Watson AI business-wide segment. One such project involves Everledger, which works with blockchain innovation to track the acquisition of expensive goods, including the exchange of precious metals. For example, IBM uses the Everledger data store for certain jewel attributes (more than 1,000,000 of them, certified by the IBM blockchain). Besides, Watson uses the information in thousands of guidelines to ensure that the jewelry is in line with UN resolutions that restrict the supply of disputed minerals. [10,11]

13.1.4 How Can Blockchain Transform AI? Redefined Intelligence

The vast majority of AI and blockchain shortcomings can be satisfactorily addressed by integrating both ecosystems [12], [13]. Man-made intelligent calculations rely on data or data to read, think, and select the worst possible goals. Machine learning computations work best when data is collected from a data repository or a reliable, secure, and large database. The blockchain fills in as a scattered record where data can be stored and processed in a cryptographically, verified, and distributed way at every point of the mining facility. Blockchain data is maintained with high integrity and flexibility, and may not be altered. When intelligent calculator systems are used to make decisions and to perform tests, the outcome of these decisions can be trusted and can be challenged. The combination of AI and blockchain can create a secure, durable, low-level structure of highly sensitive data that AI-driven systems can integrate, maintain, and implement [13].

This concept carries out the necessary mobility to ensure data and information in a variety of fields, including clinics, individuals, banks and budgets, trade, and legal data. Artificial intelligence can benefit from the discovery of various blocks of blockchain for modeling machine learning and tracking data maintained in integrated P2P frameworks. This information is usually initiated on intelligent interrelated objects that include resource collection, for example, IoT intersections, robots, large city networks, buildings, and vehicles. The features and organizations of the cloud can also be used for machine-to-machine testing and intelligent, flexible, and data-based flexibility. A few of the major features of using the AI blockchain can be summarized as follows:

- **Improved data security:** Data captured within the blockchain is much safer. Blockchains strike completely by taking care of fragile and single data in an irrational environment. Blockchain databases contain deliberately tested information, which easily affects the "private keys" which must be kept secure [14]. These are AI models that provide secure data tracking licenses and as a result, ensure reliable and reliable decision-making results.
- **Improved confidence in robotic decisions:** Any decision made by AI executives is broken when it is difficult for customers or customers to understand and trust. Blockchain has gone on strike to record trading in low-level records by the point, making it extremely difficult to accept and accept decisions taken, with the assurance that the records have not been altered, during a man-made investigation rate [14]. Recording a flexible AI system pattern in a blockchain will increase understanding and will build public trust to understand robot decisions [15]. The need for a fired analyst can be explored in a large area with a robotic climate, where the understanding of large numbers can be refined by a low-level design [14, 15, 16].
- **Collaborative decision-making:** In a climate with large robots, all the managers need to work together to achieve the set number [14, 15, 16]. Powerful images that are distributed and distributed to people have been found in various robotic systems, without the need for a central force. The robots make decisions by presenting a voting form and the results are restricted by the lion's

distribution guidelines. Each robot can live on its own as a business, where the blockchain is public to all robots. It can be used to verify the appearance of voting form results. Until the maximum number reaches the final goal, this cycle is repeated by all robots.

- **Decentralized intelligence:** By making high-level decisions involving various managers to perform sub-tasks that come with standard preparation information (e.g., in case of supervised training), some cybersecurity AI masters can be joined to provide fully structured security to basic networks and understand booking issues [17], [19].
- **High performance:** Multiuser business measures, involving a variety of delegates, for example, solitary clients, business firms, and administrative management, are naturally ineffective due to the approval of multiple business groups. The integration of AI and blockchain development empowers autonomous agents (or DAOs) with the ability to customize the acceleration and speed of data/tracking/flow of assets between different compliance [18], [20].

13.2 BLOCKCHAIN-ENABLED AI APPLICATIONS

There has been a great deal of enthusiasm to investigate blockchain developments to empower commercial institutions of various kinds. In this work, the authors have provided a blockchain creation that empowers the "man-made commercial center": a stage where consumers and data providers can create data and/or models with respect. Maintaining protection and trust during these negotiations is important. As an empowering case study, the authors have considered the exchange study space. In this setting, the shopper object needs to receive a large set of configurations from various private information providers that link to the minimum consent data provided by the customer. Information providers expect a reasonable motivation for their commitment and the consumer alike needs to maximize its profits. The authors held a dispersed conference on the blockchain that assured the safety and convenience of the shopper. Furthermore, the authors have shown that our use of the blockchain takes a critical role in tackling the problem of fair value and reliable protection.

13.2.1 BLOCKCHAIN-AI ARCHITECTURE FOR CORONAVIRUS FIGHTING

Initially, all data from clinical labs, emergency clinics, online media, and many different sources are backed up and generate raw data that forms the scale of big data. This data should be guaranteed for security and safety during the next Covid episode and testing, using blockchain. Here, the blockchain can offer a few practical responses to Covid-related management, for example, the next episode, customer safety insurance, daily safe operations, a flexible clinical chain, and the next gift. The secure data collected on the blockchain network is being investigated using intelligent AI-based programs. Using reliable forecasts and the ability to directly investigate large data collected from Covid sources, AI can provide Covid assistance with five key applications, specific research, Covid location, Covid testing, drug/vaccine development, and anticipation or future Covid similar and flare-up.

Finally, at the head of the value chain comes a layer of partners that includes circles, for example, governments, health care providers who benefit from blockchain-AI arrangements. Note that blockchain may generate secure networks and documents for setting up safer, faster, and stronger data trading with partners due to the set environment [21].

13.2.1.1 Blockchain-based Solutions for Coronavirus Fighting

In this section, the authors dismantled Covid's blockchain function against five key settings, including the following follow-up, customer safety assurance, secure daily operations, a series of flexible scenarios, and gift tracking.

- **Next outbreak:** Blockchain can provide responses that follow the Covid explosion. Without a doubt, the blockchain could understand that it provided data apparatuses in Covid's dangerous experiments. Blockchain can be viewed as distributed data records that can receive separate updates near the permanent and store them in blocks that are connected reliably and permanently. In the Covid epidemic, the blockchain may have helped thousands of people with Covid injuries by recording chronic manifestations of the disease [22].
- **User safety insurance:** While trying to control the spread of the disease, many arrangements are being considered worldwide. In the Covid epidemic, the blockchain can be used to record continuous data on indicators, locations, a significant medical problem with high protection. This can be achieved through the trust and allocation of local blockchain power.
- **Daily safe operations:** In the Covid emergency, the blockchain has been developed as a promising component that enables the administration of daily tests in virtual environments to reduce the risk of compression of the infection. Here the authors consider two common daily tasks and client-to-client tasks and shortcuts.
- **A series of therapeutic variables:** Blockchain has shown remarkable value in the use of flexible chains, for example, chain products generously, alternating chain chains [23], [24], [25], [26]. In this catastrophic emergency, maintaining a stable supply of medicines and food has become a test of the local medical services. The innovation of the blockchain could allow charitable organizations to achieve the rapid progression of change by following the stream from the sources to the opposition honestly and firmly.
- **Subsequent contribution:** The potential for blockchain in gift after applications has been tested in previous activities [27], [28]. In the case of Covid's emergency, the gift is one of the most important tests to help with the management and medical management of the injured.

13.2.1.2 AI-Based Solutions for Coronavirus Fighting

In this section, the authors presented a study of the use of AI in the fight against the Covid scourge. The artificial intelligence can support by providing the application of five principles, in particular the screening of the episode, the location of Covid, the

Covid investigation, the development of vaccines/drugs, and anticipation of future events.

- **Covid flare-up size measurement:** Performance surveillance can help combat the spread of Covid-19 by its ability to measure the size of a Covid episode by disseminating human-use projects.
- **Coronavirus detection:** Some such as late preparation using Covid detection AI have been suggested in writing. The measure is to detect heat on a person's face so that the authors can distinguish the possible indications of Covid-19 disease. Measured intelligence can integrate arrangements because of its ability to face recognition systems.
- **Conclusion and treatment of Coronavirus:** In the fight against Covid, creating effective strategies and effective treatments with astonishing ingenuity, select the achievements of combating Covid-19 infections. Measured intelligence can include responses to help to understand and combating Covid19 through intelligent data research [29], [30]. Work on [31] introduces an AI-based model, considered CovidX-Net that incorporates seven distinct formats of deep convolutional network models, for example, a modified Visual Geometry Group Network (VGG19) and a second Google MobileNet form, the Discovery of Covid-19.
- **Improving vaccine/drugs:** To finally fight the growing Covid-19 epidemic, it is necessary to build a strong immune system to fight the viral infections caused by Covid-19. Measured intelligence can come in handy as an attractive tool to aid in the formation of antibodies. Some investigators immediately suggested using AI for the project.
- **Covid-19 flare-up forecasts for the future:** Covid's continued rise includes the requirement for a future episode such as Covid-19. The types of Covid scarring and explosive control should be considered in Covid-fighting exercise modes [32], [33], [34]. Late in the day, AI was used to anticipate an episode like Covid.

13.2.2 BLOCKCHAIN AND AI TECHNOLOGY FOR NOVEL CORONAVIRUS DISEASE 2019 SELF-TESTING

The authors have suggested blockchain freedom and artificial intelligence associated with self-testing and globalization frameworks for Covid-19 and other preventable diseases. Short submissions and proper implementation of the proposed framework are likely to regulate the distribution of Covid-19 and connected deaths, particularly in areas that do not have access to the research facility infrastructure. [35]

The legend of Covid 2019 illness (Covid-19) has now reached sub-Saharan Africa (SSA) with cases reported in more than 40 SSA countries. SSA's social structures now have the effects of chronic weakness and high mortality rates linked to four times the major diseases (HIV, Tuberculosis, and Non-Communicable Diseases) weight gain [36].

Similarly, overcrowded SSA networks, unfamiliar settlements, and restricted provincial settings and assets are at risk and are often unable to protect themselves from the Covid-19 component. These people are reserved in respect of welfare administration and maybe the new Covid-19 points. The global data of Covid-19 shows surprisingly low transmission rates and low pass rates in bordered countries, especially in sub-Saharan Africa (SSA) countries.

While a small SSA community and a warm atmosphere can put SSA at the forefront of adapting to Covid-19 flare-up [37], there is growing concern about the effect of Covid-19 interactions on people living with other fragile sensitive structures, for example, HIV, TB, and diabetes as well as the framework for social welfare in restricted settings, for example, SSA nations [38,39]. There is also a growing concern about the inability to detect and report cases, especially given those with weak social structures, poor visibility, lack of research center limits, and a restricted general social base in African countries [40]. Admission to a definitive conclusion, testing, and calculation of flare-ups requires a framework for treatment resources [41]. Evidence suggests that many countries with restricted assets do not have a compelling, quick test framework [42].

These provisions further limit the accessibility of the introduction of a healthy electronic prescription for non-communicable diseases to promote resistance and control the development of undeniable diseases, for example, Covid-19 [43]. Extensive social inclusion, admission to advanced patients and appropriate patients, and the administration of research facilities (PALM) are widely expected to assist the medical care structures entrusted with achieving sustainable development goals [44]. This requires a rapid transformation of events and social development planning to ensure the accuracy and electronic monitoring of Covid-19 in unsafe areas.

Previous evidence suggests that the rapid turnaround events and planning of a targeted care objective (POC) test due to the Covid-19 component may help assess disease prevalence and reduce weight in the social framework [45,46]. The effect of rapid testing on the mortality rate of Covid-19 has been observed in Germany [47]. Growing social development, for example, the new blockchain and artificial intelligence (AI) can be integrated with POC diagnostics to enable self-assessment of patients separately due to the introduction of Covid-19. Blockchain is a computer, a public record that records online transactions. Includes computer-assisted computation of records and agreements and eliminates all mediation risks [48,49]. One of the most common uses of the blockchain is the digital currency of Bitcoin [50], which has been successfully used as a budget option in developing economies that remember SSA nations [51].

The evolution of blockchain has shown the flexibility that has recently led to a wide range of applications, including medical and medical care [52,53]. The use of blockchain and AI in medical care is evident in the following areas: record holders of electrical clinics; drugs and drugs do not fit the board; biomedical testing; education; remote patient evaluation; and the investigation of social data [54]. The practical objective of diagnostic and self-examination care has been successfully achieved in limited asset settings [55,56,57].

However, there is limited evidence on the use of blockchain and new AI to determine illness. Recalling the Covid-19 period and evidence of a framework for heavy medical services and nonprofit diagnostic frameworks in confined settings, and the use of accessible health services (mHealth), authors suggest, rapid turnaround events, and easy organization blockchain and AI-coupled mHealth with self-assessment and global positioning frameworks as one of the key response mechanisms for Covid-19 and other unavoidable diseases. The basic development of this frame-work is a mobile phone or tablet (app) that can be changed from existing testing apps [57,58,59]. The application will need a client identifier before opening pre-test references. After testing, the client will forward the results to the system.

The blockchain and AI system will enable test result trading to monitor the visual effects of all tests and continue to match the number of positive and negative test results. This will help to ensure that each of these specific cases is included as a place of isolation and treatment. A proposed blockchain for public planning and artificial intelligence combined with self-assessment associated with diversity and a global structure of growing pollution. The basic development of this structure is done with a PDA or tablet (application) that can be modified from existing testing applications [57,58,59]

13.2.3 SECURING DATA WITH BLOCKCHAIN AND AI

In this chapter, the authors have suggested SecNet: a structure that can allow secure data to be maintained, registered, and participated in the massive Internet expansion. Web access is more secure with big data verified so update AI with a larger data source. By joining three key components:

1) blockchain-based data that provides proof of ownership, including the received data participating in a large-scale space to specify large data;
2) AI-based secure acquisition platform for intelligent transmission, which helps build trust on the web;
3) Reliable about a security organization purchasing tool, provide a way to deal with people by adding financial rewards when they release their data or organization, facilitate data sharing, and appropriately benefit from the most common AI presentation.

Besides, the authors referred to the common use of SecNet in the same way as its preferred method of dealing with shipping, as well as analyzing its adequacy from network security and money-related security. [60]

SecNet is complemented by designing a more secure web, by organizing three key components:

1) blockchain-based data that provides proof of identity;
2) AI-set a secure registration category based on the details of making smart and powerful security rules;
3) the trust of the trust framework for the purchase of security organizations.

SecNet sites are affiliated with blockchain-based networking. In a network, inter-mediate points speak and appear in an understanding of blockchain strategies. At that point, they planned to make wise arrangements. Demonstrating understanding, either in the middle of a domain point or in the results of a contract release, each institutional point contains a blockchain record of the change of status at various institutional points. Specifically, SecNetCenter points have a data collection module and a data security access control module. SecNetcenters also have an operation support system (OSS) module that enables AI-based ASI (ASC) security management to deliver data and security rules from the data. [60]

To use AI and blockchain to match the problem of data management, just as you enable AI with blockchain help in trusting data on the board in unreliable weather. Authors have suggested SecNet, which is one of the few social networking sites that does not have secure data set aside, sharing and subscribing rather than disputing the offer. SecNet provides data access ensures the support of blockchain innovations, as well as the AI-based safe registration stage just as a blockchain-based promotional program, which provides global visibility and data integration capabilities and all awesome AI ultimately achieving better network security.

Besides, the authors commented on the effectiveness of the SecNet mill application in the clinical research framework and offered a selection of ways to use SecNet power. Besides, the authors reviewed its development of network vulnerabilities while combating DoS attacks, and also explored the creative component of encouraging clients to share secure network security rules. [60]

13.2.4 Blockchain Integration with Robotics and AI

In this chapter, the authors have developed a framework for various strategies and phrases that attempt to apply blockchain robustness to robotic structures, improve AI management, or address problems encountered in large blockchains, which can stimulate the ability to create robotic frameworks with increased capabilities and security. The authors presented a review, strategic review, and finalized chapter on our vision for the historic reconciliation of this technology. [61]

What the authors remove from these recommendations is that blockchain, robotics, and AI will affect the way writers live, because they can bring in such a significant amount of value without anyone else, but associating them with authors maximizes those benefits. Blockchain can store large amounts of data that can be used to prepare better AI management for robotics, however, the blockchain can also complement as part of the data communication between various robots and have smart-embedded workflows, improve robotic and inter-network performance. Unless this is undoubtedly a reality soon, existing strategies are still in their infancy, mainly since the authors are facing a period of unstable development of this development, and it is yet to be developed.

From imaginative strategies and categories, the authors have accepted that those with the most promising future are those that link multiple management on a single platform and, at the same time, share code with an open-source network and have award programs to find bugs. Indeed, the authors have seen many robotic frameworks using blockchain innovation, primarily in modern and military environments where the blockchain can assist with mechanical operations with the help of smart contracts and enable security development frameworks and cycle detection.

The blockchain introduces [61] issues of how to shed information, trust different members, and lead internal and external changes by validating data concerning the whole framework. Circumstances, when a combination of these two developments work together to reach a common goal, are not difficult to imagine. For example, dozens of "Cop Robots" look at the streets welcoming people and looking for missing habits. These robots can transcend over the blockchain and have work-related triggers. This can run when they find someone hurting someone, voting for an outline on the best plan to go to the scene, or asking for help. However, to achieve this type of behavior, smart contractors must improve security and can interact with data from outside the blockchain.

It is important to have categories that can incorporate interdisciplinary approaches so that the market deviates from different approaches to dealing with a few set-up plans, or a catastrophe is imminent, setting clear connectivity standards to enable multiple conversational responses. The commercial centers that will be seen will be important to make individual robots ready to perform a variety of complex tasks without the need for their designers to list all the important different arrangements. This can and should be combined with cloud robotics. [61]

13.2.5 BLOCKCHAIN AND DEEP REINFORCEMENT LEARNING EMPOWERED INTELLIGENT 5G BEYOND

In this chapter, the authors initially proposed protected and intelligent engineering to remotely connect remote networks by connecting AI and blockchain to remote networks to enable asset integration and compliance. At the same time, the authors have raised the issue of blockchain integration in the use of an add-on framework, and have developed another storage strategy through in-depth reinforcement study. The statistical results indicate the adequacy of the proposed structure.

The design has three aircraft:

- Cloud Plane
- Edge Plane
- Client Flight

Various technicians are equipped with powerful balances, savings, and energy management in a cloud aircraft. With a holistic view, this layer can use advanced techniques, for example, digging data and big data to move an organizational action course from an open organizational move to active organizational action, by considering a few events or allocating a couple of benefits. Due to the high subscription limit and adequate hosting resources, cloud operators can tackle applications that allow for delays and store content of large size or low reputation. Besides, there is a central controller in the cloud plane. The central officer is wearing safety gear and controls the safety limits and buttons, the items are as follows: a large scope base (MBS), a base station (MBS), road units (RSUs), telephones, and sharp vehicles. Near the layout of buildings (e.g., MBS, SBS, RSU) are placed in the camp of the organization and are staffed by MEC and blockchain.

Network networks can provide mobile and automated radio communications to achieve consistent installation and long-distance communication. MEC staff, with inventory, fixed assets, and AI capabilities, can provide remote smart and storage deployment to ensure increased counts and delay critical requests and keep the most popular or important item, for example, the latest news and warning problem at the end of the network. Blockchain can record all transactions made on a remote network and maintain a distributed record to increase security and environmental protection remotely. Trading can be a broad range, component/stock component, power exchange, and so on. [62]

The points of interest for the proposed blockchain and AI architecture, the proposed engineering are based on a variety of benefits, which are summarized as follows:

- **Safe and intelligent resource management:** In the proposed design, a list of managers among portable customers, V2V power exchange, or storage sharing can be recorded in secure squares to facilitate the distribution or distribution of secure assets. Limited intelligence statistics can naturally detect confusing remote networks, evolving management requirements, and the time-varying conditions of an accessible asset. In this way, abusing blockchain and AI, proposed engineering could create better ways to distribute assets in a secure environment.
- **Adaptable networking:** The integration of MBS, SBS, mobile, and vehicle into the proposed engineering provides an opportunity for flexible communication. Installation intelligence can accurately assess the terrain, channel activity, and current remote network restrictions, and select the most remote access mode (i.e., cell network, V2V, or D2D) to improve connectivity, reduce power consumption, or improve the client experience. For example, the design can create clear client strategies to enable some portable clients to communicate with MBS while others combine with RSUs to maintain centralized data trading.
- **Strong and dynamic orchestration:** Aided by blockchain, performance reports, and network settings that can be generated and adapted in a high-quality way between side employees, which can promote network resilience and enable reliable collaboration. Computer-based spying can create performance reports to identify and summarize other network installs and can provide a faster and more efficient system for deploying virtual machines. Besides, inventory and storage of goods on a private jet can encourage more significant network planning. [62]

In this chapter, the authors have suggested a continuous and secure structure for reducing remote networks by connecting blockchain and AI to a remote network. The proposed design can provide managers with secure and highly desirable assets, flexible communication, and reliable organization. In the meantime, the authors have introduced four common blockchains that include remote assets onboard sites, namely, rating sharing, D2D deployment, V2V power exchange, and computational uploads. Besides, the authors have misused the consortium blockchain to create a secure climate-saving object and used in-depth study reinforcement to plan a final plan to increase asset consumption. The statistical results confirm the adequacy of the proposed plan. [62]

13.2.6 A BLOCKCHAIN-ENABLED INTELLIGENT IoT ARCHITECTURE WITH AI

The authors presented an overview of the proposed architecture composed of four layers:

(1) Device intelligence,
(2) Edge intelligence,
(3) Fog intelligence, and
(4) Cloud intelligence.

Device intelligence incorporated the AI features in various IoT devices with related blockchain applications. A massive volume of data is transmitted to the smart end of the device intelligence. After that, the edges of the AI-enabled channels are connected to the blockchain. Many sensor devices are connected to each base station powered by AI in edge-edge intelligence. This is for the analysis and processing of traffic data from these hearing devices. From the ingenuity of the edge, the data used is detailed in the fog intelligence. Many AI-enabled fog sites with blockchain are integrated into fog intelligence. [63]

Every blockchain-enabled AI node is compatible with the creation of AI-equipped channels on the edge. It is imperative to process data in cloud intelligence. Finally, AI-enabled data centers are in cloud intelligence. Data centers are connected to a blockchain application to provide secure and transparent data for IoT applications like smart health care, smart transport, and much more. Besides, the entire data center announces the harvest of its data planning in the cloud intelligence management of multiple data tests solving low precision, sleep, IoT security issues. [63]

The authors proposed Block-IoT-Intelligence Architecture to transform Blockchain, IoT, and AI in the following order:

(i) **Device intelligence:** This layer is used to transform blockchain and AI for IoT applications. This layer contains many IoT devices with sensors that generate large amounts of data with the help of sensors. To keep the IoT data secure and private, the distribution method is provided via a blockchain. Rathore et al. [64] proposed a comprehensive, secure blockchain learning program in Blockchain that blockchain and distributed ledger are integrated to keep the IoT data secure across various AI systems. The blockchain network and AI are used to transfer data among the IoT devices with built-in device intelligence. The IoT device shares database-based intelligence on peer blockchain networks.

(ii) **Edge intelligence:** This layer provides the cutting edge of blockchain and AI transformation for various IoT applications. Analytical tools used by AI to find reliable data mines such as measurement, feature extraction, and representation of large unstructured data from IoT devices. In IoT applications, edge-of-the-box technology uses speech and imagery in health care, transportation, and similar applications. Blockchain technology is connected to peers on informal IoT devices to ensure the security of data. Rathore et al. [67] used blockchain and distributed ledger integration in IoT networks in a secure way of in-depth learning to reduce a single point of failure problems with high accuracy in a low-level manner. In the proposed art, the edge-of-the-box contains blockchain-based nodes connected to fog intelligence and device intelligence. Lower layer data collection and data integration are converted to fog intelligence in a distributed way. Edge intelligence addresses issues such as network expansion, resource management, and load balancing with the help of blockchain-based edge network networks.

(iii) **Fog intelligence:** This layer represents a fog concept to transform blockchain and IoT with AI. In machine learning models, AI technology has been used. They make decisions as quickly as possible in the fog of wisdom. The distributed storage location keeps a complete ledger copy of everything. By transmitting data to the IoT network in fog intelligence, general rules and regulations should be followed by all stakeholders. Rathore et al. [65] The research has been used by the authors to refine the wisdom of the fog. The secure buildup is provided by a combination of blockchain, distributed ledger, and SDN to ensure the best detection of attacks. The research work proposed by the authors measured centralization concerns with few calculations and high accuracy. The authors suggested the creation of an intelligent blockchain that allowed AI-enabled nodes to be connected via a blockchain to fog intelligence. Sharing of intermediate concerns or data is done by AI-enabled nodes to obscure intelligence, providing details of network training via blockchain. The data used is transmitted to fog intelligence from edge intelligence. In the IoT network, it is responsible for processing traffic data identification data. However, resource management, data shortages, power consumption, and problem-solving challenges are addressed with an AI-enabled node.

(iv) **Cloud intelligence:** Finally, the proposed BlockIoTIntelligence architecture used cloud intelligence to transform blockchain and IoT AI. In collecting, selecting, analyzing data from nearby locations, AI's clever agents are used in cloud intelligence. The distributed pattern is provided by blockchain for big data analytics protected on IoT. Xu et al. [66] addressed the issue of energy management issues by proposing a block-based resource management framework for decommissioning at the request of the editor. The authors referred to the proposed research work of Xu et al.'s [66] in their proposed architecture for cloud intelligence. Cloud intelligence relies on enabled data centers. With cloud surveillance, each IoT device sends its data to a cloud specialist to test large data on the network, providing high power and bandwidth communication over IoT data. Blockchain power reduction in cloud intelligence is used to install high security and high accuracy in large data testing of IoT applications.

13.2.7 APPLICATION OF BLOCKCHAIN WITH AI

In this section, the authors have given a complete review of the blockchain and artificial intelligence applications. The various application of blockchain with AI may be

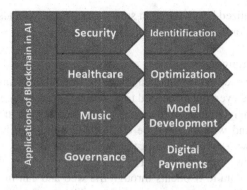

FIGURE 13.3   Application of blockchain with AI.

depicted in Figure 13.3. This decentralized blockchain system may be used in many
sectors and may transform them for better living. [68]

13.2.7.1   Global Payments

The cycle of budget exchanges on a global scale by setting up bank channels is a complex and multifaceted step involving a few go-betweens. Each step requires time and
money to measure up, which can be frustrating. Blockchain can address these issues
by providing a transfer record that keeps track of every transaction that takes place,
requires no outsiders, no intentional payments, or any deferred. When the exchange
is recorded on the record, the installment is automatically available. Installations
made for blockchain are important for organizations and consumers as this is time-
consuming, cost-effective, straightforward, and secure. Santander bank was the first
bank that empowered clients to make international installments in 2004, including
blockchain and installment applications. Artificial intelligence can be used to help a
financial business by connecting with cash with great ingenuity by resolving the most
intelligent and secure choices [69].

13.2.7.2   Blockchain Music

The music business faces some significant setbacks due to the complex structure, for
example, intellectual property rights, the direct absence of the producer, and much
more. The computer-generated music industry may overlook patents as it focuses on
arranging things. Blockchain and smart contracts create a low-level music category
that may have a specific database of the best performers or beneficiaries, enabling
a specific labeling framework that ensures sound load. Therefore, players or track
makers are carefully paid the right amount of cost as stipulated in the contracts.
Distribution of unique professional-made tracks, board distribution, optional control,
this should happen in the section designed for blockchain. For every $1000 worth of
music sold, an artist receives an average of only $23.40. As a result of the blockchain
assigned to recording, smart contracts create responsibility and understanding.
Blockchain facilitates partner work processes between smart contracts. While AI can

also be commercialized but not used in smart contracts, AI can also be used to create planning contracts or to set individual goals.

13.2.7.3 Government

The democratic framework was talked about by Republicans and Democrats in the 2016 elections, and even rallies needed to be defined. According to PC investigators, power structures may be hacked to control votes. In any case, under the record, the votes are coded and which cannot be controlled, which can be verified by independent persons whose votes have been cast. This framework similarly sets aside money for any event, the legislature. The blockchain record-level platform provides unparalleled open data to residents through the web and can make the world more than $ 2.6 trillion. These innovations can be used by many categories, for example, new companies, fish farmers, rangers, and many others as indicated by their products. Currently, once a year the data is distributed but the blockchain public record makes data available to residents at any time. With the new design of blockchain, managers can strengthen trust and responsibility by providing cybersecurity, integrated management [70].

13.2.7.4 Blockchain Identity

Blockchain verifies your personality by encoding and verifying it from spam and notifying programs. Other than that, online encounters are all about us: the few engagements we buy from donating to our human cunning in encouraging masters who submit their articles. Blockchain prevents this by creating a guaranteed data point where you just rub the data you need people to know at explicit events. At the time when a separate character is integrated, customers can bring a verified identifier such as a QR code to show their character and access specific organizations. For example, if you replace a bar, the bartender simply requires data that reveals that you are over 21 years old. Governments and organizations with computers are contributing to the blockchain trend from a distance and make it a distribution platform. Rational ingenuity has a certain level of employment in government as it will be used to further oppose social strategies, just as it helps people, when it is all over, talking to the government. [71]

13.2.7.5 Optimization

One of the highlights of the system is the artificial intelligence system to find the best order available to a given subject. Currently, the AI framework works in a variety of contexts. The AI framework shoots a cell phone, a seat, and different situations. In this way, AI editing is done so that various gadgets display different results by investigating the data. This is being done in a focused way so far but a lot of new things are being done that will improve the way it is allocated to the people. [72]

13.2.7.6 Model Development

Regular reading is one of the most important parts of the AI framework. The tone of testing and preparation is done before bringing AI applications to the market. This is done in two different ways, a cohesive and empowered approach. The installed method is robust as more data and guidance are required for preparation. Separately, the investigation of the machine apart from the other person also took place. This can be used in various programs to locate and disassemble the insect itself. [73]

13.3 CONCLUSION AND FUTURE SCOPE

In this chapter, the authors have outlined, considered, emphasize, and analyze how does blockchain related implementation concern, support or enhanced via various AI techniques. The authors have further reviewed how artificial intelligence may be exploited to achieve the goal of Blockchain 2.0. There seem to be numerous possibilities start from the current pandemic Coronavirus to blockchain music to robotics and many more in the alliance of AI and blockchain in Industry 4.0.

In this chapter, we have outlined that blockchain for AI applications are still in its immaturity, and there are still a big path to travel in terms of many research challenges which needs to be addressed in areas like privacy, smart contract security, trusted oracles, scalability, consensus protocols, standardization, interoperability, quantum computing resiliency, and governance.

REFERENCES

1. Atzori, L., Iera, A. and Morabito, G. 2010. The Internet of Things: A Survey. *Computer Networks*, 54: 2787–2805. http://dx.doi.org/10.1016/j.comnet.2010. 05.010.
2. Dogo, E. M., Salami, A. F., Aigbavboa, C., and Nkonyana, T. 2019. Taking Cloud Computing to the Extreme Edge. In a review of "Mist Computing for Smart Cities and Industry 4.0 in Africa" in *Edge Computing*, Springer, Cham, pp. 107–32.
3. Sharma, P. K., Chen, M. Y., Park, J. H. 2017. A Software-Defined Fog Node Based Distributed Blockchain Cloud Architecture for IoT, *IEEE* Access, 6: 115–24, http://dx.doi.org/10.1109/ACCESS.2017.2757955.
4. Rathore, S., Sharma, P. K., and Park, J. H. 2017. XSSClassifier: An efficient XSS Attack Detection Approach Based on Machine Learning Classifier on SNSs, *J. Inf. Process. Syst.*, 13(4).
5. Gupta, H., Vahid Dastjerdi, A., Ghosh, S. K., and Bhavya, R. 2017. iFogSim: A Toolkit for Modeling and Simulation of Resource Management Techniques in the Internet of Things, Edge, and Fog Computing Environments, *Software Pract. Exp.*, 47(9): 1275–96. http://dx.doi.org/10.1002/spe.2509.
6. Salah, K., Rehman, M. H. U., Nizamuddin, N., and Al-Fuqaha, A. 2017. Blockchain for AI: Review and Open Research Challenges, *IEEE* Access, 7: 10127–49, http://dx.doi. org/10.1109/ACCESS.2018.2890507.
7. Sundmaeker, H., Guillemin, P., Friess, P., and Woelfflé, S. 2010. Vision and Challenges for Realizing the Internet of Things. A Cluster of European Research Projects on the Internet of Things, *Eur. Comm.* 3(3): 34–6.
8. Hartman, J. H., Murdock, I., and Spalink, T. 1999. The Swarm Scalable Storage System. In *Proceedings 19th IEEE International Conference on Distributed Computing Systems (Cat. No. 99CB37003)*, pp. 74–81. https://doi.org/10.1109/ICDCS.1999.776508.
9. Yuan, Y., and Wang, F. Y. 2016. Towards Blockchain-Based Intelligent Transportation Systems. In *2016 IEEE 19th International Conference on Intelligent Transportation Systems, ITSC, IEEE*, pp. 2663–8. http://dx.doi.org/10.1109/ITSC.2016.7795984.

10. Rabah, K. 2016a. Overview of Blockchain as the Engine of the 4th Industrial Revolution. *Mara Res. J. Bus. Manag.*, 1(1): 125–35.

11. Rabah, K. 2016b. Digital Cryptoeconomics Powered by Digital Cryptocurrency. *Mara Res. J. Comput. Sci. Inf. Secur.*, 1(1): 107–31. ISSN 2518-8453.

12. Panarello, A., Tapas, N., Merlino, G., Longo, F., and Puliafito, A. 2018. Blockchain and IoT Integration: A Systematic Survey. *Sensors*, 18: 2575.

13. Marwala, T., and Xing, B. 2018. Blockchain and Artificial Intelligence. https://arxiv. org/abs/1802.04451.

14. Marr, B. 2018. Artificial Intelligence and Blockchain: 3 Major Benefits of Combining These Two Mega-Trends. www.forbes.com/sites/bernardmarr/2018/03/02/artificial-intelligenceand-blockchain-3-major-benefits-of-combining-these-two-mega-trends/.

15. Campbell, D. 2018. Combining AI and Blockchain to Push Frontiers in Healthcare. www.macadamian.com/2018/03/16/combining-ai-and-blockchain-in-healthcare.

16. Ferrer, E. C. 2016. The Blockchain: A New Framework for Robotic Swarm Systems. https://arxiv.org/abs/1608.00695.

17. Brambilla, M., Ferrante, E., Birattari, M., and Dorigo, M. 2013. Swarm Robotics: A Review from the Swarm Engineering Perspective. *Swarm Intell.*, 7(1): 1–41.

18. Strobel, V., Ferrer, E. C., and Dorigo, M. 2018. Managing Byzantine Robots via Blockchain Technology in a Swarm Robotics Collective Decision Making Scenario. In *Proc. 17th Int. Conf. Auto. Agents MultiAgent Syst. International Foundation for Autonomous Agents and Multiagent Systems: Stockholm*, Sweden, pp. 541–9.

19. Janson, S., Merkle, D., and Middendorf, M. 2008. A Decentralization Approach for Swarm Intelligence Algorithms in Networks Applied to Multi Swarm PSO, *Int. J. Intell. Comput. Cybern.*, 1(1): 25–45.

20. Magazzeni, D., McBurney, P., and Nash, W. 2017. Validation and Verification of Smart Contracts: A Research Agenda. *Computer*, 50(9): 50–7.

21. Novikov, S. P., Kazakov, O. D., Kulagina, N. A., and Azarenko, N. Y. 2018. Blockchain and Smart Contracts in a Decentralized Health Infrastructure. In *2018 IEEE International Conference Quality Management, Transport and Information Security, Information Technologies (IT and QM and IS)*, pp. 697–703.

22. Blockchain in Healthcare: the Case of Coronavirus. www.e-zigurat.com/innovation-school/blog/blockchain-inhealthcare/.

23. Juma, H., Shaalan, K., and Kamel, I. 2019. A Survey on Using Blockchain in Trade Supply Chain Solutions, *IEEE Access*, 7: 184115–32.

24. Gonczol, P., Katsikouli, P., Herskind, L., and Dragoni, N. 2018. Blockchain Implementations and Use Cases for Supply Chains – a Survey, *IEEE Access*, 8(11): 856–71.

25. Kouhizadeh, M., and Sarkis, J. 2018. Blockchain Practices, Potentials, and Perspectives in Greening Supply Chains, *Sustainability*, 10(10): 3652.

26. Wu, H., Cao, J., Yang, Y., Tung, C. L., Jiang, S., Tang, B., Liu, Y., Wang, X., and Deng, Y. 2019. Data Management in Supply Chain Using Blockchain: Challenges and a Case Study. In *2019 28th International Conference on Computer Communication and Networks (ICCCN)*, pp. 1–8.

27. Sirisha, N. S., Agarwal, T., Monde, R., Yadav, R., and Hande, R. 2019. Proposed Solution for Trackable Donations Using Blockchain. In *2019 International Conference on Nascent Technologies in Engineering (ICNTE)*, pp. 1–5.

28. Saleh, H., Avdoshin, S., and Dzhonov, A. 2019. Platform for Tracking Donations of Charitable Foundations Based on Blockchain Technology. In *2019 Actual Problems of Systems and Software Engineering (APSSE)*, pp. 182–7.

29. Chiusano, M. L. 2020. The Modelling of Covid19 Pathways Sheds Light on Mechanisms, Opportunities and on Controversial Interpretations of Medical Treatments. v2, arXiv preprint arXiv:2003.11614.

30. Basiri, M. R. 2020. Theory about Treatments and Morbidity Prevention of Corona Virus Disease (Covid-19), *Journal of Pharmacy and Pharmacology*, 8: 89–90.

31. Hemdan, E. E. D., Shouman, M. A., and Karar, M. E. 2020. Covidx-net: A Framework of Deep Learning Classifiers to Diagnose Covid-19 in X-Ray Images, arXiv preprint arXiv:2003.11055.

32. Nadim, S. S., Ghosh, I., and Chattopadhyay, J. 2020. Short-Term Predictions and Prevention Strategies for Covid-2019: A Model Based Study, arXiv preprint arXiv:2003.08150.

33. Vattay, G. 2020. Predicting the Ultimate Outcome of the Covid-19 Outbreak in Italy, arXiv preprint arXiv:2003.07912.

34. Fanelli, D., and Piazza, F. 2020. Analysis and Forecast of Covid-19 Spreading in China, Italy and France, *Chaos, Solitons and Fractals*, 134: 109761.

35. Mashamba-Thompson, T. P., and Crayton, E. D. 2020. Blockchain and Artificial Intelligence Technology for Novel Coronavirus Disease 2019 Self-Testing. Diagnostics, 10(4): 198.

36. Institute for Health Metrics and Evaluation (IHME). 2018. Findings from the Global Burden of Disease Study 2017; IHME: Seattle, WA, USA. www.healthdata.org/sites/default/files/files/policy_report/2019/GBD_2017_ Booklet. pdf (accessed May 25, 2020).

37. Chopera, D., 2020. Can Africa Withstand COVID-19? www.project-syndicate.org/commentary/africa-covid19-advantages-disadvantages-by-denis-chopera-2020-03-2020-03 (accessed May 25, 2020).

38. Wong, E. 2020. TB HIV and COVID-19: Urgent Questions as Three Epidemics Collide. https://theconversation.com/tb-hiv-and-covid-19-urgent-questions-as-threeepidemics-collide134554?fbclid=IwAR3ycjutsVRKxcRjxcsO4VawyyKf16Gey3GTMeAejvZVs ACcf9 Cgq HP0Q (accessed May 25, 2020).

39. Loven, C. N. 2020. On-Again, Off-Again Looks to Be Best Social-Distancing Option. https://news.harvard.edu/gazette/story/2020/03/how-to-prevent-overwhelming-hospitals-and-build-immunity/ (accessed May 28, 2020).

40. Whiteside, A. 2020. Covid-19 Watch: The Crisis Deepens. https://alan-whiteside.com/2020/03/25/covid-19-watch-the-crisis-deepens-2/?fbclid=IwAR3RkSOESoQzkxT4kKsIAYmDOPG9t39qoCUYQxmVLg-tnBmMtB6Zm-FpuTY (accessed March 25, 2020).

41. Herida, M., Dervaux, B., Desenclos, J.-C. 2016. Economic Evaluations of Public Health Surveillance Systems: A Systematic Review. *Eur. J. Public Health*, 26: 674–80.

42. Rattanaumpawan, P., Boonyasiri, A., Vong, S., and Thamlikitkul, V. 2018. Systematic Review of Electronic Surveillance of Infectious Diseases with Emphasis on Antimicrobial Resistance Surveillance in Resource-Limited Settings. *Am. J. Infect. Control*, 46: 139–46.

43. United Nations. 2019. The Sustainable Development Goals Report 2019. https://unstats.un.org/sdgs/report/2019/The-Sustainable-Development-Goals-Report-2019.pdf (accessed May 25, 2020).

44. Pang, J., Wang, M. X., Ang, I. Y. H., Tan, S. H. X., Lewis, R. F., Chen, J. I. P., Gutierrez, R. A., Gwee, S. X. W., Chua, P. E. Y., and Yang, Q. 2020. Potential Rapid Diagnostics, Vaccine and Therapeutics for 2019 Novel Coronavirus (2019-nCoV): A Systematic Review. *J. Clin. Med.*, 9: 623.

45. Wang, C., Horby, P. W., Hayden, F. G., and Gao, G. F. 2020. A Novel Coronavirus Outbreak of Global Health Concern. *Lancet*, 395: 470–3.
46. U.S. News. Experts: Rapid Testing Helps Explain Few German Virus Deaths. 2020. www.usnews.com/news/health-news/articles/2020-03-09/experts-rapid-testing-helps-explainfew-german-virus-deaths.
47. Yaqoob, S., Khan, M., Talib, R., Butt, A., Saleem, S., Arif, F., and Nadeem, A. 2019. Use of Blockchain in Healthcare: A Systematic Literature Review. *Int. J. Adv. Computer Science Application*, 10: 644–53.
48. Gomez, M., Bustamante, P., Weiss, M. B., Murtazashvili, I., Madison, M. J., Law, W., Mylovanov, T., Bodon, H., and Krishnamurthy, P. 2020. Is Blockchain the Next Step in the Evolution Chain of [Market] Intermediaries? https://ssrn.com/abstract=3427506.
49. Nakamoto, S. n.d. Bitcoin. A Peer-To-Peer Electronic Cash System. Available online: https://bitcoin.org/bitcoin.pdf (accessed May 25, 2020).
50. Vincent, O., and Evans, O. 2019. Can Cryptocurrency, Mobile Phones, and Internet Herald Sustainable Financial Sector Development in Emerging Markets? *Journal of Transaction Management*, 24: 259–79.
51. Mettler, M. 2016. Blockchain Technology in Healthcare: The Revolution Starts Here. In *Proceedings of the 2016 IEEE 18th International Conference on E-Health Networking, Applications and Services (Healthcom), Munich, Germany*, 14, 16, pp. 1–3.
52. Agbo, C. C., Mahmoud, Q. H., and Eklund, J. M. 2019. Blockchain Technology in Healthcare: A Systematic Review. *Healthcare*, 7: 56.
53. Zhang, P., Schmidt, D. C., White, J., and Lenz, G. 2018. Blockchain Technology Use Cases in Healthcare. In *Advances in Computers*; Elsevier, New York, vol. 111, pp. 1–41.
54. Makhudu, S. J., Kuupiel, D., Gwala, N., and Mashamba-Thompson, T. P. 2019. The Use of Patient Self-Testing in Low-and Middle-Income Countries: A Systematic Scoping Review. *Point Care*, 18: 9–16.
55. Bervell, B., and Al-Samarraie, H. 2019. A Comparative Review of Mobile Health and Electronic Health Utilization in Sub-Saharan African Countries. *Soc. Sci. Med.*, 232: 1–16.
56. Adeagbo, O., Kim, H.-Y., Tanser, F., Xulu, S., Dlamini, N., Gumede, V., Mathenjwa, T., Bärnighausen, T., McGrath, N., and Blandford, A. 2020. Acceptability of a Tablet-Based Application to Support Early HIV Testing among Men in Rural KwaZulu-Natal, South Africa: A Mixed Method Study. *AIDS Care*, pp. 1–8.
57. Pai, N. P., Behlim, T., Abrahams, L., Vadnais, C., Shivkumar, S., Pillay, S., Binder, A., Deli-Houssein, R., Engel, N., and Joseph, L. 2013. Will an Unsupervised Self-Testing Strategy for HIV Work in Health Care Workers of South Africa? A Cross Sectional Pilot Feasibility Study, 8, e79772.
58. Tay, I., Garland, S., Gorelik, A., and Wark, J. D. 2017. Development and Testing of a Mobile Phone App for Self-Monitoring of Calcium Intake in Young Women. *JMIR mHealth and uHealth*, 5: e27.
59. Kuupiel, D., Bawontuo, V., and Mashamba-Thompson, T.P. 2017. Improving the Accessibility and Efficiency of Point-of-Care Diagnostics Services in Low-and Middle-Income Countries: Lean and Agile Supply Chain Management. *Diagnostics*, 7: 58.
60. Wang, K., Dong, J., Wang, Y., and Yin, H. 2019. Securing Data with Blockchain and AI. *IEEE Access*, 7: 77981–77989. DOI: 10.1109/ACCESS.2019.2921555.
61. Lopes, V., and Alexandre, L. A. 2019. An Overview of Blockchain Integration with Robotics and Artificial Intelligence. *Ledger*, 4. https://doi.org/10.5195/ledger.2019.171
62. Dai, Y., Xu, D., Maharjan, S., Chen, Z., He, Q., and Zhang, Y. 2019. Blockchain and Deep Reinforcement Learning Empowered Intelligent 5G Beyond, *IEEE Network*, 33(3): 10–17. DOI: 10.1109/MNET.2019.1800376.

63. Singh, S. K., Rathore, S., and Park, J. H. 2020. BlockIoTIntelligence: A Blockchain-Enabled Intelligent IoT Architecture with Artificial Intelligence. *Future Generation Computer Systems*, 110: 721–43. doi: 10.1016/j.future.2019.09.002.

64. Rathore, S., Pan, Y., and Park, J. H. 2019. BlockDeepNet: A Blockchain-Based Secure Deep Learning for IoT Network. *Sustainability* , 11: 3974. http://dx.doi.org/10.3390/11143974.

65. Rathore, S., Kwon, B. W., and Park, J. H. 2019. BlockSecIoTNet: Blockchain-Based Decentralized Security Architecture for IoT Network, *Journal of Netwoks Computer Application*. http://dx.doi.org/10.1016/j.jnca.2019.06.019.

66. Xu, C., Wang, K., and Guo, M. 2017. Intelligent Resource Management in Blockchain Based Cloud Data Centers, *IEEE Cloud Computer*, 4(6): 50–9. http://dx.doi.org/10.1109 /MCC .2018.1081060.

67. Rathore, S., and Park, J. H. 2019. DeepBlockIoTNet: A Secure Deep Learning Approach with Blockchain for the IoT Network, *Trans. Ind. Inform*, 11(14): 3974. DOI: https://doi.org/10.3390/su11143974.

68. Gulati, P., Sharma, A., Bhasin, K., and Azad, C. 2020. Approaches of Blockchain with AI: Challenges and Future Direction. *Proceedings of the International Conference on Innovative Computing and Communications (ICICC) 2020*, Available at SSRN: https://ssrn.com /abstract= 360073 5http://dx.doi.org/10.2139/ssrn.3600735.

69. Goel, A. K., Rose, A., Gaur, J., and Bhushan, B. 2019. Attacks, Countermeasures and Security Paradigms in IoT. *2019 2nd International Conference on Intelligent Computing, Instrumentation and Control Technologies (ICICICT)*. doi: 10.1109/icicict46008.2019. 8993338.

70. Lin, Y., Lin, Y., and Liu, C., 2019. AItalk: A Tutorial to Implement AI as IoT Devices. *IET Networks*, 8(3): 195–202.

71. Zhang, G., Li, T., Li, Y., Hui, P., and Jin, D. 2018. Blockchain-Based Data Sharing System for AI-Powered Network Operations, *Journal of Communications and Information Networks*, 3(3): 1–8.

72. Arora, D., Gautham, S., Gupta, H., and Bhushan, B. 2019. Blockchain-Based Security Solutions to Preserve Data Privacy and Integrity. *2019 International Conference on Computing, Communication, and Intelligent Systems (ICCCIS)*. doi: 10.1109/icccis48478.2019.8974503.

73. Marr, B. 2020. Artificial Intelligence and Blockchain: 3 Major Benefits Of Combining These Two Mega-Trends, Forbes.

14 Financial Portfolio Optimization

An AI Based Decision-Making Approach

Mansi Gupta, Sonali Semwal, and Shivani Bali

CONTENTS

14.1 INTRODUCTION

Portfolio management is not only an art but also science of making decisions about the investment mix and policy, matching investments to objectives of investors, asset allocation for individuals and institutions, and stabilizing risk against performance. Financial market is surrounded by inconsistent and complex factors which are determined by endogenous and exogenous factors (Filimonov and Sornette 2012). In managing a portfolio one should consider the volatility and the correlation corresponding to the various financial instruments. AI has been a part of the whole sector. In the field of finance AI and ML can be use in numerous ways such as risk assessment, fraud detection and management, financial advisory services, and trading, managing finance. The use of AI-powered mechanism helps in generating accurate and reliable data. Most digital banks and loan issuing apps make use of machine learning algorithms.

The core competence of banks is the management of credit risk as no other customer has superior knowledge and experience in relation to the management of credit risk of the customer. Earlier algorithm for portfolio management comprises the bit of machine learning elements but recently advance deep learning has been used to resolve the problem of portfolio management. The use of technology such as Excel, R, and Python makes things easier. In the same way ML and AI has also made complicated financial techniques easy (Mangram and Myles 2013). AI and deep learning have made potential replacement of certain roles of human beings in the field of finance. Financial data is a large volume data which has a very large amount of data available for training because of which financial data is suitable for deep learning (Goodfellow, Bengio, and Courville 2016). With advancement in deep learning and machine learning it has become possible to estimate the complex behavior of financial instruments and to automatize the decision-making process up to certain extent Jangmin, Lee, Lee, and Zhang 2006). Prediction of stock market behavior was first introduced by Atsalakis and Valavanis in 2009. With the development of quantitative strategies, the asset management industry has tremendously changed in the last few years. Machine learning methods and artificial intelligence have shaped the investment processes (Perrin and Roncalli 2019). AI and ML have also made it easier for investor to make comparative analysis. A comparative performance analysis between the portfolio optimization strategies can be undertaken which is based on various performance measures such as portfolio expected return, standard deviation, Beta coefficient, Sharpe Ratio, Jensen's alpha (Jason Narsoo 2017).

An integrated approach which considers risk and return equally could provide in-depth knowledge about portfolio structure. Modern Portfolio Theory is a mathematical framework which assembles portfolio of assets in such a manner that investors who are risk averse can construct a portfolio to maximize return based on the market risk level. MPT was founded by the economist Harry Markowitz in 1950s. The basic principle of Markowitz was that there should be a proper balance between risk and return. MPT uses the weights for returns of equities to calculate the portfolio-wide return. Investors are more focused on evaluating the risk and return of individual securities while constructing a portfolio (Ivanova and Dospatliev 2017). An investor can engineer a portfolio by considering alpha coefficient and beta coefficient, both the factors determine the performance of a portfolio. Alpha coefficient measures an investment performance relative to its risk, whereas beta measures the investment's return relative to the market (Omisore, Yusuf, and Christopher 2012). The objective is to find a combination of weights which assists in finding out the portfolio with highest Sharpe Ratio as it compares the performance of an investment with a risk-free asset, after adjusting for its risk.

The Markowitz Model was one of the brilliant innovations in the science of portfolio selection. By taking variance as a measure of risk, he suggested a method of selecting the right portfolio which provides maximum returns for a given level of risk (Vinay Kumar 2018).

14.2 LITERATURE REVIEW

According to Klotz and Lindermeir (2015), authors find the cluster analysis suitable method to determine various multivariate contract specifications based on the dataset provided by the financial institutions. The purpose was to improvise the decision-making process in credit portfolio management with the help of data mining method.

Mokta Rani Sarkar (2013) conducted a study on the Dhaka Stock Exchange in Bangladesh. The objective of the study was to conduct a risk-return analysis of individual securities listed in DSE. Also, the author assisted the investors in selection process to make the right choice and constructed the optimal portfolio using the Markowitz Model. The chapter concluded that risk and return plays a crucial role in making the investment decisions. Though the model was complex and time consuming, but the model performed well in DSE.

Obeidat, Shapiro, Lemay, MacPherson, and Bolic (2018) create a system which assists in automating portfolio management and improving risk adjusted returns. The authors created a system which estimates the returns and recommends the portfolio for investments. The study concluded that LSTM could provide a better risk adjusted returns than conventional strategic passive portfolio management approach. Liang, Chen, Zhu, Jiang, and Li (2018) used three different algorithms, deep deterministic policy gradient (DDPG), proximal policy optimization (PPO) and policy gradient (PG) in portfolio management. The purpose was to show that PPO is more appealing than the other two algorithms. The authors observed that deep reinforcement learning capture patterns of market movements even though it is allowed to observe the limited data. Also, deep reinforcement learning does not accomplish such remarkable performance as those in games and robot control.

According to Bailey and Prado (2013), most of the financial practitioners encountered a problem which is portfolio optimization. The aim of their paper is to fill a gap by providing a well-documented, step-by-step open-source implementation of critical line algorithm (CLA) in scientific language. Isaac Siwale (2013) used two generic models for portfolio optimization problem. This proprietary solver is known as GENO. Siwale's paper reviewed the portfolio model of Harry Markowitz and its objective was to present the conceptual methods which effectively address the issues which apparently impede its implementation in practice. Vijayalakshmi Pai (2019) has explained the use of Python in the field of specialization, especially portfolio optimization in seven strides, from fundamentals to constrained portfolio optimization models.

The paper by Glensk and Madlener (2015) provides certain examples to illustrate the application of financial models for the portfolio optimization of power generation assets. These financial models help in determining the optimal time to invest.

Joshi and Parmar (2020) carried out a research to test whether the Markowitz Framework of Construction of Portfolio offers the improved investment alternative to investors of the Indian stock market. The model was applied to 30 listed securities on the Bombay stock exchange. It was concluded that the model provides better insights to the investors for the investment purpose. Soleymani and Paquet (2020) tested the performance of Deep-Breath framework with four test sets over three distinct investment periods. The authors found efficiency in the approach and found that deep reinforcement learning is more suitable for long-term investment.

14.3 METHODOLOGY

A portfolio is a collection of financial investments like commodities, bonds etc. A key to earn reasonable returns from a portfolio is diversification which means not to put all your eggs in one basket.

A sample portfolio consisting of four stocks, i.e., Hindustan Unilever, Reliance Industries, Tata Consultancy Services (TCS), and Sun Pharma from four different industries was constructed. Data was extracted from Yahoo Finance from January 1,

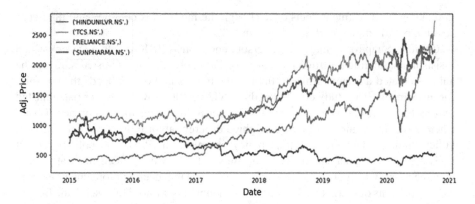

FIGURE 14.1 Adjusted closing price.

2015 to October 7, 2020 using pandas data reader function and looked into the correlation between the stocks.

Portfolio performance have been measured using portfolio return, risk, and Sharpe ratio

$$\text{Portfolio Return} = w1*r1 + w2*r2 + ...+wn*rn$$

W is the weight assigned to security and r is the return from that security.

Returns were annualized by multiplying by 250 as on average there are 250 trading days in a year.

$$\text{Portfolio Risk} = \sqrt{\left(x_1\ x_2\ ...x_n\right)\cdot\begin{pmatrix} \sigma_1^2 & \cdots & \sigma_{1n} \\ \vdots & \ddots & \vdots \\ \sigma_{n1} & \cdots & \sigma_n^2 \end{pmatrix}\begin{pmatrix} x_1 \\ \vdots \\ x_n \end{pmatrix}}$$

Where X is the weight assigned to each security and n is total number of stocks

Modern Portfolio Theory has been followed to optimize the portfolio, which was propounded by an American economist Harry Markowitz. The theory is all about the maximizing the return investors could earn on their investment portfolio considering all the risks associated to it. Markowitz said that investors can make a portfolio to maximize the returns by accepting a quantifiable amount of risk. Statistical analysis has been used for the measurement of risk and for selection of assets in a portfolio in an efficient manner we have used the mathematical programming.

Sharpe Ratio was proposed by Nobel laureate William F. Sharpe and it is performed to help investors. The ratio compares the return of an investment and the risk associated to the investment.

$$\text{Sharpe Ratio} = \frac{Rp - Rf}{\sigma p}$$

Rp and Rf represents the return of portfolio and risk-free rate, respectively.

The difference between the risk and return is divided by the standard deviation of the portfolio's excess return (σp). The ratio basically calculates the risk-adjusted return. It helps the investors to understand that whether portfolio's excess returns are because of investment decision which is a smart decision or a result of too much risk. The higher the Sharpe Ratio, the better the risk-adjusted-performance. Risk free rate is taken as 7 percent based on the rates offered on government securities.

10,000 different portfolios have been simulated for the stocks, and then annual return, variance and Sharpe Ratio have been calculated for each portfolio. Portfolio with highest Sharpe Ratio and lowest variance (risk) has then been identified. We have used different packages in Python to design an optimal portfolio such as matplotlib and seaborn for visualization, numpy for mathematical operations, pandas for data manipulation and analysis, datetime module for supplies of classes to the work with date and time.

At the end, the program tells how many stocks of each company to buy based on investor's budget and weights assigned.

14.4 ANALYSIS

The return plot (Figure 14.2) shows that return over the five-year span has been same for each company considered. There have been some ups and downs and those are majorly accounted by the financial results declared by the company.

The Markowitz plot (Figure 14.3) shows the return, risk, and sharpe ratio of each portfolio. The Sharpe Ratio measures the risk-adjusted return. Risk-adjusted returns determine the return earned by an investment after considering the degree of risk. A portfolio is considered to be superior if it has Higher Sharpe Ratio. Highest Sharpe Ratio is higlighted with the diamond (1) while the minimum risk is highlighted with diamond (2) among which an investor can make a choice based on his risk and return preferences.

A comparative study of equal weighted portfolio, portfolio under maximum Sharpe ratio and minimum risk is given in Table 14.1.

It is clear from the table that return is maximum when we opt for the maximum Sharpe Ratio portfolio, i.e., 28 percent while it is only 18.7 percent in case of equally weighted portfolio and 19.5 percent in case when the risk is minimum. There is not much of a difference between risk under 3 approach as compared to return.

So, by simulating portfolios and following Markowitz theory, an investor can maximize its return while the risk does not increase much.

The first and foremost aim of a portfolio is to maximize the return with the given level of risk. So, the investor always tries to maintain a tradeoff between the risk and return. Weights assigned to each security when risk is minimum is shown in Figure 14.4. The maximum weight is assigned to Hindustan Unilever (0.373) followed by TCS (0.349), Sun Pharma (0.150), and Reliance (0.127).

Sharpe Ratio considers the standard deviation to calculate the risk-adjusted returns. It can be used to compare the risk-adjusted returns across all the categories of financial instruments. The ratio gives an idea to the investor of how much extra return

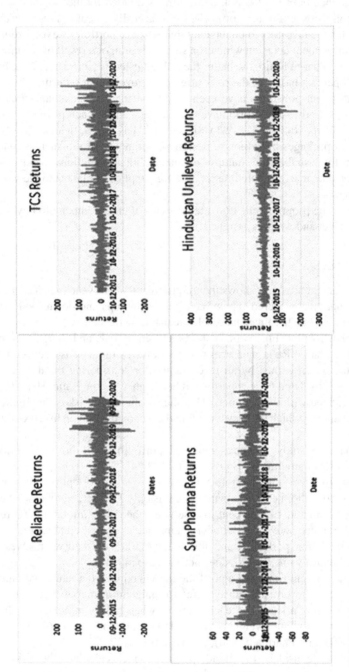

FIGURE 14.2   Return from each stock.

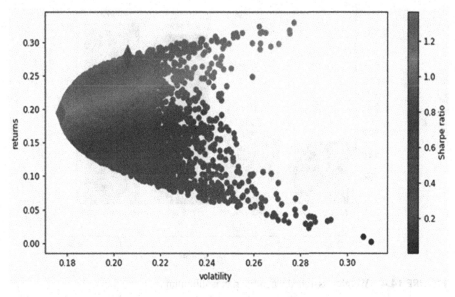

FIGURE 14.3   Markowitz plot return, risk, and Sharpe Ratio.

TABLE 14.1
Comparative analysis under different approaches

Metrics	Equally weighted Portfolio	Maximum Sharpe Ratio	Minimum Risk
Return	0.187	0.28	0.195
Risk	0.185	0.206	0.177
Sharpe Ratio	0.633	1.017	0.703

he/she may be earning in the volatile market for holding an asset with risk. Weights assigned to each security when Sharpe Ratio is maximum are shown in Figure 14.5. The maximum weight is assigned to Reliance (0.484) followed by Hindustan Unilever (0.340), TCS (0.349), and Sun Pharma (0.150)

14.5   CONCLUSION AND FUTURE SCOPE

Investing and portfolio management is part art and part science. Optimal portfolio construction is a major objective of an investor. During the process of constructing an optimal portfolio we consider several factors such as risk and return of the individual assets. Correlation among individual assets along with their risk and return are helpful as it is an important determinant of portfolio risk. To design an optimized portfolio is important for an investor to have an understanding with the risk. The main purpose of portfolio creation is to diversify risk and maximize return. The chapter aimed to do that by following Markowitz theory and simulating numerous portfolios. Markowitz theory works under an assumption that an investor is risk averse which means that

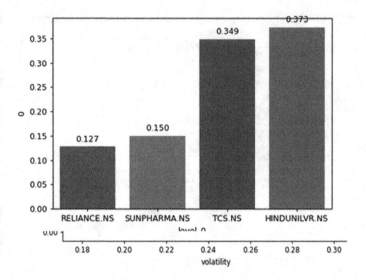

FIGURE 14.4 Weights assigned when the risk is minimum.

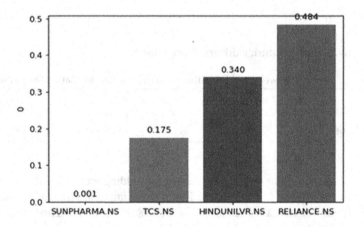

FIGURE 14.5 Weights assigned when Sharpe Ratio is maximum.

investors prefer a portfolio with less risk. It was found return can be maximized and risk can be minimized by simulating multiple portfolios. Using Python for portfolio creation makes the process quicker and easier. It helps in automating the process and gives more efficient results than manual calculation.

The chapter followed Markowitz theory that suggests return can be maximized for a given level of risk and risk can be minimized for given level of return using the Sharpe Ratio, which is one of the most important tools to determine the performance of any fund and investment. This ratio may help the investor to select the right financial instrument to construct an optimized portfolio. Only four stocks were considered in the sample portfolio; risk can be further minimized by either adding more securities or replacing with some other securities. Risk free rate has been taken at 7 percent, and the Sharpe Ratio can change with the risk-free rate as well. Results could further be improved by using neural network and deep learning models and considering a strategic Markowitz approach.

AI and ML have revolutionized the different sectors in the global economy. Adoption of such advanced tools and techniques have changed the face and the operations of the industry. Banking and financial sector are making immense use of such tools to keep track of data. In the field of finance, it is normal to deal with the high volume of data and accurate historical records for which ML and AI is the most suited tool. Such advancement has provided positive in the sector results such as improved customer support, enhanced data quality, fraud prevention, digital financial assistants, and customized marketing strategies.

REFERENCES

Bailey, D. H. and Prado, M. L. 2013. An Open Source Implementation of the Critical-Line Algorithm for Portfolio Optimization, *Algorithms* 6(1): 169–96.

Filimonov, V. and Sornette, D. 2012. Quantifying Reflexivity in Financial Markets: Towards a Prediction of Flash Crashes. *Phys. Rev. E.* 85. 10.1103/PhysRevE.85.056108.

Glensk, B. and Madlener, R. 2015. Review of Selected Methods for Portfolio Optimization of and Irreversible Investment in Power Generation Asset Under Uncertainty, *Ekonometria*, 4: 20–42.

Ivanova, M. and Dospatliev, L. 2017. Application of Markowitz Portfolio Optimization on Bulgarian Stock Market from 2013 to 2016, *International Journal of Pure and Applied Mathematics*, 117(2): 291–307..

Iyiola, O., Munirat, Y., and Nwufo, C. 2012. The Modern Portfolio Theory as an Investment Decision Tool. *Journal of Accounting and Taxation*, 4(2).

Jangmin, O., Lee, J., Lee, J. W., and Zhang, B.-T. 2006. Adaptive Stock Trading with Dynamic Asset Allocation Using Reinforcement Learning. *Inf. Sci.*, 176: 2121–47. 10.1016/j.ins.2005.10.009.

Klotz, S. and Lindermeir, A. 2015. Multivariate Credit Portfolio Management Using Cluster Analysis, *Journal of Risk Finance*, 16(2): 145–63.

Krishna, J. and Parmar, C. 2020. Application of Markowitz Model in Indian Stock Market Reference to Bombay Stock Exchange, *Multidisciplinary International Research Journal of Gujarat Technological University*, 2.

Kumar, V. 2018. A Simplified Perspective of the Markowitz Portfolio Theory. *International Journal of Research and Analytical Reviews*, 5(3): 193–6.

Liang, Z., Chen, H., Zhu, J., Jiang, K., and Li, Y. 2018. Adversarial Deep Reinforcement Learning in Portfolio Management, *Journal of Risk Finance*, 17(4): 153–67.

Mangram, M. 2013. A Simplified Perspective of the Markowitz Portfolio Theory. *Global Journal of Business Research*, 7.

Narsoo, J. 2017. Performance Analysis of Portfolio Optimisation Strategies: Evidence from the Exchange Market. *International Journal of Economics and Finance*, 9: 124. 10.5539/ijef.v9n6p124.

Obeidat, S., Shapiro, D., Lemay, M., MacPherson, M. K., and Bolc, M. 2018. Adaptive Portfolio Asset Allocation Optimization with Deep Learning, *International Journal on Advances in Intelligent System*, 11(1–2).

Perrin, S. and Roncalli, T. 2019. Machine Learning Optimization Algorithms and Portfolio Allocation. 10.13140/RG.2.2.13566.95047.

Sarker, M. R. 2013. Markowitz Portfolio Model: Evidence from Dhaka Stock Exchange in Bangladesh, *IOSR Journal of Business and Management*, 8(6): 68–73.

Siwale, I. 2013. Practical Portfolio Optimization. 1. 10.13140/RG.2.1.2491.2883/1.

Soleymani, F. and Paquet, E. 2020. Financial Portfolio Optimization with Online Deep Reinforcement Learning and Restricted Stacked Autoencoder – Deep Breath, *Expert Systems with Applications*, 156: 113456.

Vijayalakshmi Pai, G. A. (2019). Python for Portfolio Optimization: The Ascent! https://www.researchgate.net/publication/343904773_Python_for_Portfolio_Optimization_-The_Ascent_G_A_Vijayalakshmi_Pai.

15 Intelligent Framework and Metrics for Assessment of Smart Cities

Krishna Kant Sharma, Arvind Maurya, and Reema Thareja

CONTENTS

15.1 INTRODUCTION

Ever since humans settled down in habitation, the population has usually consisted of two different centers; one which indulged in agriculture and subsistence activity and the other one which indulged in economic and other activities. As more and more people gravitated towards urban centers of living, the first wave of urbanization occurred in human history.

The Indus Valley civilization which can be tracked as far back as 3300 BC witnessed a sophisticated level of urbanization. This is expressed in many ways; size, architectural elaboration, some sense of town planning, city layout, water acquisition, and management (Possehl 2002). The remarkable town planning is reflected in the elaborate and intricate network of roads, houses, granaries, drainage system, and great baths. This is the earliest mention of urbanization in Asia during the Bronze Age and offers insights to modern day town planners as well.

At the turn of the century, advancement in agriculture and industrial revolution resulted in large scale migration to urban centers. As per a study conducted by the United Nations, it is estimated that the world population grew from 1.6 billion in 1900 to over 6 billion in 2000 and is projected to reach 10 billion by 2050 (United Nations 2019).

Rural flight or migration from rural areas to urban centers of development for better prospects is putting all the available resources of a city under tremendous pressure. This is particularly true for fast-growing economies like India and China. In Asia, Tokyo, Osaka, Beijing, Shanghai, Guangzhou, Mumbai, Kolkata, Delhi, Dhaka, Karachi, Jakarta are amongst the most heavily populated urban cities (United Nations 2019).

This large-scale movement of people to cities puts a tremendous stress on all the resources including food supply, healthcare, housing, sanitation, transport, etc. which lead to air and water pollution, crime, and social deprivation.

Urban Heat Islands too are a direct consequence of rapid urbanization characterized by densely populated urban areas which are lot hotter than surrounding rural areas. Air and water pollution from various sources is directly responsible for this phenomenon.

In 2020, 1100 scientists from 153 countries released a statement in which they noted that the world was facing a serious climate emergency and sustainable development was the only way forward and to achieve that would require a significant change in which our society interacts with natural ecosystems (Ripple, et al. 2020).

Post Second World War there was a flurry of unbridled construction across the world. In 1972, the United Nations Conference on Human Environment in Stockholm raised its concerns about natural resources and environment. However, realizing that several of the key goals had not been met, the World Commission on Environment and Development (WCED) in its report "Our Common Future" in 1987 stressed on the importance of sustainable development to ensure a better world for the future. The mission released its report "Our Common Future" in 1987 and defined sustainable development. According to the Commission, sustainable development is one that meets the needs of the present without compromising the ability of future generations to meet their own needs (World Commission on Environment and Development, 1987).

15.2 REVIEW OF EXISTING LITERATURE

Rapid development in information and communication technology (ICT) has helped city planners devise new concepts of developing cities.

In 1994, the Dutch termed their initiative of connecting households over the Internet network in Amsterdam as De Digitale Stad or The Digital City. The idea was expanded by Graham and Aurigi (1997) who illustrated how access to Internet will empower and ensure greater participation of local population and the emergence of virtual cities. In 1999, Dubai announced the launch of Dubai Internet City which intended to use digital transformation using ICT and enhance governance in the city.

TABLE 15.1

Smart city, sustainable city indicators

ISO Standard	Year	Indicators
37120	2014	Indicators for city services and quality of life
37122	2019	Indicators for smart cities
37123	2019	Indicators for resilient cities

Though "smart city" is the most widely used term, other terms such as "digital city," "virtual city," "connected city," and "resilient city" are also widely used by companies and researchers.

Smart city, sustainable city indicators as mentioned by the International Organization for Standardization (ISO) are given in Table 15.1.

The ISO/ITU standards focused primarily on the key indicators to measure smart city transformation.

In 2014 Andrea Zanella, Nicola Bui, Angelo Castellani, Lorenzo Vangelista, and Michele Zorzi introduced the idea of an Urban urban IoT system which would exploit the benefits of ICT to provide value added services for smart city (Zanella et al. 2014).

In 2015, the IEEE Smart City Group laid stress on how Internet of Things (IoT) using sensors, networks, data analytics is fundamental to smart city development.

More recently, the ISO/IEC 21972:2020 standard focuses on upper level ontology for smart city indicators. The standard defines a data model that can be used to represent smart city indicator definitions which in turn reflect the role played by ICT in smart decision making.

Traditional approach towards developing frameworks for smart cities focused on the SMELTS approach (Social, Management, Legal, Economic, Sustainability) with smartphones being the driving force behind this (Joshi et al. 2016). Other approaches to measure and analyze the impact of smart solutions in pursuit of sustainable development by smart cities introduced terms such as smartainability (Girardi et al. 2016). However, the ever-pervasive technology also leads to new challenges. It was found that existing laws were simply not enough to safeguard the private data collected by smart city infrastructure (Yang et al. 2018). Sustainable Collaborative Network as a result of collaboration between government agencies and citizens were considered robust, flexible, and efficient (Yahia et al. 2019).

However, most of the traditional literature looks at most facets of smart cities in silos and do not aggregate them to provide a comprehensive framework which would encapsulate the true dynamics of a smart city.

Till now, not even a single universally accepted definition of smart city is available. Smart city means different things to different people across the world depending upon the resources available, level of development, and aspirations of citizens. Also, the recent Covid-19 pandemic has laid bare the inadequacies of the existing city infrastructure and has also served as a learning experience for town planners. Different wings of civic agencies can no longer operate in silos; there has to be coordination and synergy between city agencies to provide services to its citizens. The reaction time to any event also needs to be faster for a meaningful response.

15.3 DEFINITION OF PROPOSED ASSESSMENT FRAMEWORK

Sustainable development aided by rapid developments in technology has incentivized city planners to design and develop futuristic cities which meets the social, physical, professional, safety, and aspirational needs of its citizens.

We have designed a similar framework to assess any smart city and what can be the potential future action items to make sure that it meets the needs of futuristic challenges. The framework can be used by city planners and administrators to develop a more holistic way of developing/designing a smart city. IoT, with its vast array of interconnected sensors collecting data which is then sent to cloud servers for analytics and decision making, is central to this framework. This framework meets the challenges forced by technology by turning them into enablers. It also meets the sustainable aspects of the city.

The six critical pillars of the framework as identified by us are:

1. Municipal corporation
2. Safety and security
3. ICT infrastructure
4. Interconnectivity
5. Sustainability
6. Critical event management

Below we give a brief description of the evaluation parameters for each of these six critical pillars.

15.3.1 MUNICIPAL CORPORATION

Municipal corporation is generally the first point of interaction of the citizens with the government. Most of their needs, requirements, services, amenities etc are provided by the municipal corporations.

Water, electricity, gas, traffic and tele-education are some of the areas where technology can be integrated with the needs of the people to provide for a better living. Table 15.2 lists evaluation parameters and their corresponding critical factors which are necessary to analyze a smart city project.

15.3.2 SAFETY AND SECURITY

Safety needs are identified as the basic needs of any individual. They are needed to be fulfilled before the higher psychological needs are to be met. Video monitoring of crime, fire safety and compliance, cyber security are the critical factors which need to be addressed by city planners. Table 15.3 lists some of the parameters that are used to assess a smart city with security perspective.

TABLE 15.2
Municipal corporation evaluation parameters and critical factors

Evaluation parameter	Critical factors
Smart energy meters and management	Smart meters for energy monitoring, billing, payment, issues identification, customer issues redressal, availability of electricity hours, predictive power shutdowns, energy need forecasting and infra planning
Intelligent traffic management	Analytics driven traffic light control, monitoring traffic violation and ticket issuance, ensuring tickets closure with fine, escalation mechanism for areas with regular and more traffic violations, citizen alerts of road constructions and blockage
Smart water metering and billing	Clean safe drinking water availability, leakage identification, predictive maintenance
Smart gas metering and billing	Self-billing, analytics-based demand prediction, query resolution system, analytics based infra issue identification and planning
Tele-education	Connectivity in the schools, training of teachers for tele-education, digital material, Secure online exams

TABLE 15.3
Safety and security evaluation parameters and critical factors

Evaluation parameter	Critical factors
Video crime monitoring	Deployment of video surveillance, drone-based video monitoring of sensitive areas, video analytics, and alerting
Fire safety and compliance	Monitoring of safety measurements in public places and big complexes, analytics driven compliance
Cyber security	Protection of government systems for cyber-attack, establishment of cells for cybercrime redressal, infrastructure to investigate cyber security issues

15.3.3 ICT INFRASTRUCTURE

The Information and communication infrastructure in the city should be affordable and available to every citizen. The infrastructure should also be robust enough to withstand and cater to any evolving demand. The recent onslaught of the Covid-19 pandemic, which necessitated long lockdowns of the city, tracking and treating infected persons, and long durations of work-from-home situation, put an enormous strain on the existing infrastructure and laid bare the inefficiencies in the existing scenario. The incident has also served as lessons as to what measures are needed to be taken to ensure an infrastructure which doesn't collapses in any emergency. Table 15.4 lists some parameters that can be used to assess a smart city based on its infrastructural set-up.

TABLE 15.4
ICT evaluation parameters and critical factors

Evaluation parameter	Critical factors
Availability	Reach of information and communication technology to every citizen
Quality	Quality of infrastructure
Standardization	Non-standardized and non-interoperable information, communication, and technology (ICT) solutions suffer from constraints and will pose a risk to smart city projects

TABLE 15.5
Interconnectivity evaluation parameters and critical factors

Evaluation parameter	Critical factors
Connected systems	Seamless interconnectivity of different government departments to make citizen's experience smooth
Command center	Providing visibility of performance of different departments, option to issue remedial action in case of any problem, tracking of remedial action to closure
Integrated multi-modal transport	Providing interconnected system for transport with single-payment system

15.3.4 INTERCONNECTIVITY

One of the biggest challenges that the city planners face is coping with the numerous standards and protocols that govern the functioning of various ICT equipment. The lack of standardization means that most equipment doesn't "talk" to each other and there is wastage and cost over runs.

A central command and control center is at the heart of any smart city as it interconnects various departments, services, etc. to give an aggregated view of city functioning. Some key indicators for evaluating a smart city based on its interconnectivity are listed in Table 15.5.

15.3.5 SUSTAINABILITY

Sustainability remains at the core of any smart city. Efficient and optimum use of resources and their equitable distribution to every citizen is one of the key drivers of smart city.

Waste management, carbon control governed and managed by analytics driven infrastructure planning is a step towards achieving this goal. Some crucial parameters and their critical factors are summarized in Table 15.6.

TABLE 15.6
Sustainability evaluation parameters and critical factors

Evaluation parameter	Critical factors
Waste management	Sensor driven waster systems for pick up, systems to convert waste into energy/fuel, compost etc. wastewater treatment plants, waste control system to measure effectiveness and take corrective actions across the board
Carbon control	Green initiatives, measurement for effectiveness and corrective actions
Analytics driven infrastructure planning	Data mining to forecast infrastructure requirements

TABLE 15.7
Critical event management evaluation parameters and critical factors

Evaluation parameter	Critical factors
Pandemic preparedness	Systems to assess pandemic situations and communication to citizens
Riot management	Video analytics-based riot predictions, connectivity to different departments like police, fire, medical to alert, monitoring equipped vehicles for live assessment
Communication systems	Systems to deliver message directly to citizens in critical events

15.4 CRITICAL EVENT MANAGEMENT

The recent Covid-19 pandemic has exposed all the inefficiencies and shortcomings of our cities. Our framework lays stress that the smart cities of future should be geared up to meet the challenges of any future pandemic. Riot management and communication systems are two other critical factors which need the attention of city planners. Other key factors are listed in Table 15.7.

15.5 ASSESSMENT FRAMEWORK

The 20 evaluation parameters have been divided into five different maturity levels, i.e., 2, 4, 6, 8, 10. All of these five maturity levels have detailed measurable indices based upon the preparedness of the city across these indices. Weights have been assigned to each of these 20 evaluation parameters with detailed performance evaluation. The city under review is assessed across all of these evaluation parameters and then the score is summed up to arrive at the final status of the city under review. The final score gives the visibility of the city's preparedness and also the key areas where it needs to improve to move up the rankings. Several cities can be assessed at the same time using this framework at the same time and the differences between them

TABLE 15.8
Evaluating assessment framework

Evaluation parameter	Weightage
Smart meters and management	5%
Intelligent traffic management	5%
Smart water metering and billing	5%
Smart gas metering and billing	5%
Tele-education	5%
Video crime monitoring	5%
Fire safety and compliance	5%
Cyber security	5%
ICT infra availability	5%
ICT infra quality	5%
Standardization	5%
Connected systems	5%
Command center	5%
Integrated multi-modal transport	5%
Waste management	5%
Carbon control	5%
Analytics driven infra planning	5%
Pandemic preparedness	5%
Riot management	5%
Communication Systems	5%
Score (out of 100)	

will give the city planners and administrators an insight into the livability and sustainability of any city. Some crucial assessment factors for assessing a smart city are summarized in Table 15.8.

15.6   RATING CRITERIA

The rating criteria to assess smart city framework for all the key evaluation parameters have been detailed in Table 15.9.

TABLE 15.9
Rating criteria

S.No	Evaluation parameter / ranking criteria	Weights	Range
1	Smart meters and management – energy	2	0–20% households are equipped with smart meters and basic management system in place
		4	21–40% households are equipped with smart meters and basic management + digital issues addressal system is in place
		6	41–60% households are equipped with smart meters and basic management + digital issues redressal + predictive power shutdown system in place
		8	61–80% households are equipped with smart meters and basic management + digital issues redressal + predictive power shutdown + energy need forecasting system in place
		10	81–100% households are equipped with smart meters and advanced digitalization in place including analytics based infra planning
2	Intelligent traffic management	2	0–20% traffic lights are equipped with analytics driven control
		4	21–40% traffic lights are equipped with analytics driven control + automated system for violation monitor
		6	41–60% traffic lights are equipped with analytics driven control + automated system for violation monitor + automated violation resolutions
		8	61–80% traffic lights are equipped with analytics driven control + automated system for violation monitor + automated violation resolutions
		10	81–100% traffic lights are equipped with analytics driven control + automated system for violation monitor + automated violation resolutions + advanced road alert systems
3	Smart water metering and billing	2	0–20% households are equipped with smart meters and basic management system in place
		4	21–40% households are equipped with smart meters and basic management + digital issues addressal system is in place

(continued)

TABLE 15.9 (Continued)
Rating criteria

S.No	Evaluation parameter / ranking criteria	Weights	Range
		6	41–60% households are equipped with smart meters and basic management + digital issues redressal + leakage identification systems in place
		8	61–80% households are equipped with smart meters and basic management + digital issues redressal + leakage identification system + predictive maintenance system is place
		10	81–100% households are equipped with smart meters and basic management + digital issues redressal + leakage identification system + predictive maintenance system is place + analytics driven planning to meet clean safe drinking water needs
4	Smart gas metering and billing	2	0–20% households are equipped with smart meters and basic management system in place
		4	21–40% households are equipped with smart meters and basic management + digital issues addressal system is in place
		6	41–60% households are equipped with smart meters and basic management + digital issues redressal +self services options to optimize opex cost
		8	61–80% households are equipped with smart meters and basic management + digital issues redressal +self services options to optimize opex cost + analytics driven fulfillment system in place
		10	81–100% households are equipped with smart meters and basic management + digital issues redressal +self services options to optimize opex cost + analytics driven fulfillment system in place
5	Tele-education	2	0–50% schools are equipped with connectivity and school management system in place for attendance, results, updates etc.
		4	51–100% schools are equipped with connectivity and school management system in place for attendance, results, updates etc.

TABLE 15.9 (Continued)
Rating criteria

S.No	Evaluation parameter / ranking criteria	Weights	Range
		6	100% schools are equipped with connectivity and school management system in place for attendance, results, updates etc. + readiness of centralized infrastructure for offline teaching
		8	100% schools are equipped with connectivity and school management system in place for attendance, results, updates etc. + readiness of centralized infrastructure for offline teaching + availability of digital material for teaching
		10	100% schools are equipped with connectivity and school management system in place for attendance, results, updates etc. + readiness of centralized infrastructure for offline teaching + availability of digital material for teaching + digital teachers training
6	Video crime monitoring	2	0–50% area covered for video surveillance with crime detection capabilities
		4	51–100% area covered for video surveillance with crime detection capabilities
		6	51–100% area covered for video surveillance with crime detection capabilities + drone based surveillance monitoring of sensitive areas
		8	51–100% area covered for video surveillance with crime detection capabilities + drone based surveillance monitoring of sensitive areas+ crime database management systems
		10	51–100% area covered for video surveillance with crime detection capabilities + drone based surveillance monitoring of sensitive areas+ crime database management systems + system for video analytics based case resolutions
7	Fire safety and compliance	2	0–40% coverage and automated monitoring of buildings to comply with fire safety and security

(continued)

TABLE 15.9 (Continued)
Rating criteria

S.No	Evaluation parameter / ranking criteria	Weights	Range
		4	41–80% coverage and automated monitoring of buildings to comply with fire safety and security
		6	81–100% coverage and automated monitoring of buildings to comply with fire safety and security
		8	0–50% coverage of connectivity with fire stations and analytics driven compliance
		10	51–100% coverage of connectivity with fire stations and analytics driven compliance
8	Cyber security	2	0–50% government establishments are protected and externally audited for cyber-attacks
		4	51–100% government establishments are protected and externally audited for cyber-attacks
		6	Capacity to cover up to 50% citizens for cybercrime redressals supported by evidence of cases reported, time to resolve, resolution provided, satisfaction from citizens
		8	Capacity to cover up to 100% citizens for cybercrime redressals supported by evidence of cases reported, time to resolve, resolution provided, satisfaction from citizens
		10	Advanced analytics driven systems in place for analyzing cybercrime issues with directed optimization targets for optimizations of earlier defined metrices
9	ICT infra availability	2	0–20% households are covered with ICT Infra Availability
		4	21–40% households are covered with ICT Infra Availability
		6	41–60% households are covered with ICT Infra Availability
		8	61–80% households are covered with ICT Infra Availability
		10	81–100% households are covered with ICT Infra Availability

TABLE 15.9 (Continued)
Rating criteria

S.No	Evaluation parameter / ranking criteria	Weights	Range
10	ICT infra quality	2	Definition of metrics are in place to assess quality of ICT Infra
		4	Automated monitoring is in place to measure defined metrices
		6	Policies definition in place to enforce compliance of metrics and take corrective actions
		8	Transparent systems in place on metrices being measured, their status, corrective actions identified, status
		10	Advanced analytic based quality improvement systems are in place
11	ICT infra standardization	2	Standardization are defined and measurement pre-requisites are in place
		4	0–30% compliance level
		6	31–70% compliance level
		8	71–100% compliance level
		10	Iterative processes in place to refine standardization
12	Connected systems	2	All the government departments working in silos
		4	0–30% departments are connected to share relevant data and digitalized
		6	31–70% departments are connected to share relevant data and digitalized
		8	71–100% departments are connected to share relevant data and digitalized
		10	Single window redressal of citizens issues irrespective of department involved in it
13	Command center	2	0–30% departments are covered with systems and technology in place for layered visibility of functioning of different departments to key stakeholders like chief minister, secretary etc.
		4	31–70% departments are covered with systems and technology in place for layered visibility of functioning of different departments to key stakeholders like chief minister, secretary etc.

(continued)

TABLE 15.9 (Continued)
Rating criteria

S.No	Evaluation parameter / ranking criteria	Weights	Range
		6	31–70% departments are covered with systems and technology in place for layered visibility of functioning of different departments to key stakeholders like chief minister, secretary etc.
		8	The data from different sources and city systems is available to be easily aggregated together to gain far greater insight into what is going on in the city
		10	Predictive and forecasting systems to upfront identification of possible challenges in operating different departments and recommended actions
14	Integrated multi-modal transport	2	No integration of transport options available for citizens like buses, metros, trains etc.
		4	NA
		6	Partial integration of transport modes available
		8	NA
		10	Single payment, multi-mode transport option available with digital payment options
15	Waste management	2	0–30% areas are covered with sensor driven waste bins to optimize effective pickup
		4	31–70% areas are covered with sensor driven waste bins to optimize effective pickup
		6	71–100% areas are covered with sensor driven waste bins to optimize effective pickup
		8	Effective supply chain of wastage treatment to covert waste into energy/fuel, compost etc.
		10	Analytics driven waste control system to measure effectiveness and take corrective actions across the board
16	Carbon control	2	Systems in place to measure current carbon emission from city and factors attributing to it like energy distribution, transport system, street lighting etc.

TABLE 15.9 (Continued)
Rating criteria

S.No	Evaluation parameter / ranking criteria	Weights	Range
		4	Recurring go green initiatives and targeted reduction in carbon emission by 20%
		6	Recurring go green initiatives and targeted reduction in carbon emission by 40%
		8	Recurring go green initiatives and targeted reduction in carbon emission by 80%
		10	Funded research initiates are in place to keep continuous improvement in carbon emission
17	Analytics driven infra planning	2	Identification of departments required for infra planning and requisite measurement systems deployment to capture data
		4	0–30% departments onboarded with required technologies to capture data
		6	31–70% departments onboarded with required technologies to capture data
		8	71–100% departments onboarded with required technologies to capture data
		10	Analytics driven system to forecast needs of infrastructure based on collated data for next 5 year, 10 years and 20 years
18	Pandemic preparedness	2	Systems and technology in place for responding to pandemic situation like disease tracing, up to date and correct information availability, geolocation tracking etc.
		4	Smart supply chain setup with more automation and less human interventions
		6	Systems in place for wireless inspection of civic violations and response plans to counter the violations
		8	Hospitals are equipped with smart intensive case units and isolation wards which can be activated if required.
		10	Controlled autonomous delivery system in place in coordination with essential services delivery organizations
19	Riot management	2	Riot response team definition and training program identified/used which can be triggered in case of riots
		4	Infrastructure in place in form of vehicles equipped with live monitoring and connectivity. Number is predicted through historical data

(continued)

TABLE 15.9 (Continued)
Rating criteria

S.No	Evaluation parameter / ranking criteria	Weights	Range
		6	Analytics driven assessment using social media data during any event to predict any unforeseen happening
		8	Riot specific video/data analysis systems to predict any possible riot based on feeds from other systems like video crime monitoring, Cyber Security etc.
		10	Analytics driven feedback to planning team to develop smooth access to areas during riots and preventive measures identification
20	Communication systems	2	Communication systems are in place to deliver critical messages directory to citizens
		4	Contacts capture of citizens and update process is in place
		6	Systems are integrated to get risk alerts from different authorized agencies and decision workflow is in place
		8	Metrics are defined and measured to assess effectiveness of communication system
		10	Checks and balances and required policies are in place to prevent misuse of the system

15.7 CONCLUSIONS

Rising population and the strain on resources coupled with unforeseen challenges mean that city planners and administrators have their hands full. Smart cities should also ensure that there is no digital divide amongst its citizens. They should be reliable, scalable, and more efficient than cities of before.

As smart cities gain popularity across the world, it is essential that global standards, protocols, definitions are crystallized to design, develop, and measure any smart city across a host of parameters. The recent Covid-19 pandemic has been a learning incident for everyone as it has laid bare the sheer inadequacy in our cities in dealing and coping with any unexpected calamity.

Smart cities require new skills and competencies. They will need to be networked, connected, and collaborative. The rapid advancements in technology act as the perfect leverage for creating smarter cities and empower citizens. IoT with its inherent features will be the engine that will power cities to be smarter, safer, and sustainable.

REFERENCES

Girardi, P., and Temporelli, A. 2016. Smartainability: A Methodology for Assessing the Sustainability of the Smart City. https://www.sciencedirect.com/science/article/pii/S187661021730276X.

Joshi, S., Saxena, S., Godbole, T., and Shreya (2016) Developing Smart Cities: An Integrated Framework. https://www.sciencedirect.com/science/article/pii/S0268401219302452.

Our Common Future. 1987. Report of the World Commission on Environment and Development. https://sustainabledevelopment.un.org/content/documents/5987our-common-future.pdf.

Possehl, G. L. 2002. The Indus Civilization: A Contemporary Perspective. https://books.google.co.in/books/about/The_Indus_Civilization.html?id=pmAuAsi4ePIC.

Ripple, W. J., Wolf, C., Newsome, T. M., Barnard, P., and Moomaw, W. R. 2020. World Scientists' Warning of a Climate Emergency. https://academic.oup.com/bioscience/article/70/1/8/5610806.

United Nations. 2019. World Population Prospects – Population Division. https://population.un.org/wpp/.

Yahia, N. B., Eljaoued, W., Saoud, N. B. B., and Colomo-Palacios, R. 2019. Towards Sustainable Collaborative Networks for Smart Cities Co-governance. https://www.sciencedirect.com/science/article/pii/S0268401219302452.

Yang, F. and Xu, J. 2018. Privacy Concerns in China's Smart City Campaign: The Deficit of China's Cybersecurity Law. https://onlinelibrary.wiley.com/doi/full/10.1002/app5.246.

Zanella, Z., Bui, N., Castellani, A., Vangelista, L., and Zorzi, M. 2014. Internet of Things for Smart Cities. https://ieeexplore.ieee.org/document/6740844.

Index

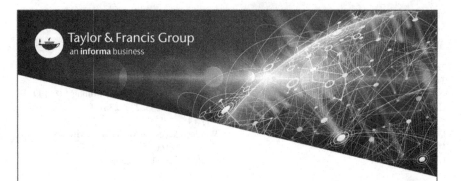

Printed in the United States
by Baker & Taylor Publisher Services